Find Your Way Around JCT 98

Private With Quantities

T0186748

Richard Woolley

London and New York

First published 2001 by
Spon Press, 11 New Fetter Lane, London, EC4P 4EE.
Simultaneously published in the USA and Canada by
Spon Press, 29 West 35th Street, New York, NY 10001.

Spon Press is an imprint of the Taylor & Francis Group

© Richard Woolley

Printed and bound in Great Britain by TJ International Ltd, Padstow, Cornwall

Publisher's Note
This book was prepared from camera-ready copy supplied by the author

British Library Cataloguing in Publication Data
A catalogue record for this book is available from the British Library

Library of Congress Cataloging in Publication Data
A catalog record for this book has been requested

ISBN   0-415-23482-4

CONTENTS

CONTENTS

CONTENTS

THE PURPOSE OF THIS BOOK

Although they are legal documents, Contract Forms should be seen primarily as sets of practical rules whose function is to provide smooth and efficient systems for establishing and controlling responsibilities amongst the participants on matters related mainly to quality, performance and payment. For them to achieve their purpose satisfactorily, all participants must know and understand their duties, responsibilities and undertakings thoroughly.

By their nature, construction contracts are complex documents in which reference to a particular topic covered by the Conditions may be found in several Clauses in addition to the Clause principally applicable to that topic. To read and retain permanently in the memory, not only all the provisions of such a document as JCT 98 but also their precise location in it, is a feat of which few would claim to be capable. Nonetheless, the ability to find quickly in the Conditions all the requirements related to a particular topic, is one of the fundamental duties of each participant whether he be the employer, contractor, architect, engineer, quantity surveyor or lawyer.

This essentially practical book has been compiled to assist you find all the references to each of over seventy of the topics most commonly encountered and discussed in the administration of JCT 98.

HOW TO USE THIS BOOK

In the first place, it must be understood that this book is a topic index. It does not attempt to interpret or comment on the Contract Conditions in any way. Anything contained in it, must not be read as an opinion or view on any matter covered by the Contract. The brief descriptions given under the heading 'Signpost' simply indicate the subjects which will be found covered by the particular Clause or Sub-Clause within the context of the topic heading at the top of the page.

As an example to illustrate the use of this book, consider there is a decision to be made concerning the Completion Date. Turn to the pages headed **COMPLETION DATE.** The first item on the first page shows you where to look in the Conditions to find the main Clause dealing with the topic. The other 19 items listed are the locations in the Conditions where you will find further reference to the topic of COMPLETION DATE. The descriptions under 'Signpost' will tell you whether the subject mentioned within each of those other Clauses is likely to be relevant to the particular problem before you. In every case, you must read the Contract after being directed by the 'Signpost' .

The descriptions given as 'Signposts' only indicate the subject matter of the references. They are not to help you avoid reading the Contract.

THE APPENDIX

Topics listed in the Appendix are also frequently the subject of debate but have not been included in the main body of the book as reference is made to each of them within no more than two clauses.

THE SCOPE OF THIS BOOK

This publication analyses the Articles of Agreement, Parts 1 – 5 and the Appendices and includes Amendment 1 (June 1999), Amendment 2 (January 2000) and Optional Clause 30.4A (Januaty 2000) for the 'PrivateWith Quantities' Variant of the Standard Form of Building Contract 1998 Edition issued by the Joint Contracts Tribunal.

Sundry abbreviations

CCMGD	Certificate of Completion of Making Good Defects
CDM	Construction (Design and Management) Regulations 1994
CIMAR	Construction Industry Model Arbitration Rules
cl.	Clause
cls.	Clauses
CPC	Certificate of Practical Completion
DLP	Defects Liability Period
H&S	Health and Safety
NSC	Nominated Sub-Contractor
N. Supp.	Nominated Supplier
PC	Practical Completion
PS	Price Statement
PSW	Performance Specified Work
QS	Quantity Surveyor
RICS	Royal Institution of Chartered Surveyors
SMM7	Standard Method of Measurement 7th Edition
VAT	Value added tax

ABBREVIATIONS

Abbreviated Clause Headings

Art.	Article
Articles	Articles of Agreement

PART 1   Part 1: General

Definitions etc.	Interpretation, definitions etc. (cl.1)
Contract Sum/Adjustments	Contract Sum - additions or deductions - adjustment - Interim Certificates (cl.3)
Contract Documents	Contract documents - other documents - issue of certificates (cl.5)
Statutory obligations	Statutory obligations, notices, fees and charges (cl.6)
Levels and setting out	Levels and setting out of the Works (cl.7)
Royalties etc.	Royalties and patent rights (cl.9)
Access to the Works	Access for Architect to the Works (cl.11)
Variations/Prov. Sums	Variations and provisional sums (cl.13)
Contractor's Price Statement	Alternative A: Contractor's Price Statement (cl.13.4.1.2)
Variation instruction	Variation instruction – Contractor's quotation in compliance with the instruction (cl.13A)
Value added tax	Value added tax - supplemental provisions (cl.15)
Mats. and goods unfixed	Materials and goods unfixed or off-site (cl.16)
Practical Compl./Defects	Practical completion and Defects Liability (cl.17)

Partial Possession   Partial possession by Employer (cl.18)

Abbreviated Clause Headings

Assignment /Sub-Contracts	Assignment and Sub-Contracts (cl.19)
Injury and indemnity	Injury to persons and property and indemnity to Employer (cl.20)
Insurance: persons etc.	Insurance against injury to persons and property (cl.21)
All Risks – Contractor	Erection of new buildings – All Risks Insurance of the Works by the Contractor (cl.22A)
All Risks – Employer	Erection of new buildings – All Risks Insurance of the Works by the Employer (cl.22B)
Insurance - existing structures:	Insurance of existing structures – insurance of Works in or extensions to existing structures (cl.22C)
Insurance: loss of LADs	Insurance for Employer's loss of liquidated damages (cl.22D)
Joint Fire Code	Joint Fire Code - compliance
Possession, completion	Date of Possession, completion and postponement (cl.23)
Damages	Damages for non-completion (cl.24)
Loss and Expense	Loss and expense caused by matters materially affecting regular progress of the Works (cl.26)
Determination by either	Determination by Employer or Contractor (cl.28A)
Works by Employer	Works by Employer or persons employed or engaged by Employer (cl.29)
Certificates/payments	Certificates and payments (cl.30)
CIS	Construction Industry Scheme (CIS) (Cl.31)

ABBREVIATIONS

Abbreviated Clause Headings

PART 2 — Part 2: Nominated Sub-Contractors and Nominated Suppliers

General — General (cl.35.1 - 2)

Procedure — Procedure for nomination of a sub-contractor (cls.35.3 - 9)

Payment: — Payment of nominated sub-contractor (cl.35.13)

Extension of Period — Extension of period or periods for completion of nominated sub-contract works (cl.35.14)

Failure to complete — Failure to complete nominated sub-contract works (cl.35.15)

Practical Completion — Practical completion of nominated sub-contract works (cl.35.16)

Early Final Payment — Early final payment of nominated sub-contractors (cls. 35.17-19)

Position of Employer — Position of employer in relation to nominated subcontractor (cl.35.20)

Clause 2.1 of NSC/W — Clause 2.1 of agreement NSC/W - position of contractor (cl.35.21)

Re-nomination — Circumstances where re-nomination necessary (cl.35.24)

Determination — Determination or determination of employment of nominated sub-contractor - architect's instructions (cls.35.25-26)

PART 3 — Part 3: Fluctuations

Fluctuations – taxes etc. — Contribution, levy and tax fluctuations (cls.38.1 – .2)

Sub-let/Domestics	Fluctuations – work sub-let – Domestic Sub-Contractors (cl.38.3)
Clause 38 provisions	Provisions relating to clause 38 (cls.38.4 - .6)

Abbreviated Clause Headings

Fluctuations – cost, taxes etc.	Labour and materials cost and tax fluctuations (cls.39.1 - .3)
Sub-let/Domestics	Fluctuations – work sub-let – Domestic Sub-Contractors (cl.39.4)
Clause 39 provisions	Provisions relating to clause 39 (cls.39.5 - .7)
Price adjustment formulae	Use of price adjustment formulae

PART 4	Part 4: Settlement of disputes – adjudication – arbitration – legal proceedings
PART 5	Part 5: Performance Specified Work

APPENDICES

Code of Practice: Clause 8.4.4	Code of Practice: referred to in clause 8.4.4
Advance Payment Bond	Annex 1 to Appendix: Terms of Bonds Advance Payment Bond
Off-site Materials Bond	Annex 1 to Appendix: Terms of Bonds Bond in respect of payment for off-site materials and/or goods
VAT Agreement	Supplemental Provisions (the VAT Agreement)
EDI	Annex 2 to the Conditions: Supplemental Provisions for EDI
Retention Bond	Optional clause 30.4A Contractor's bond in lieu of Retention

FOREWORD

It gives me great pleasure to introduce a second edition of Richard Woolley's book the aim of which is to provide an easy route through the JCT Form of Contract.

Since the last edition, time in the building and construction world has definitely not stood still. The JCT itself has come under the spotlight. In paragraph 5.1 of his final report *Constructing the Team*, Sir Michael Latham noted that most building work is undertaken under main contracts provided by the Joint Contracts Tribunal (JCT) though these are frequently amended. In paragraph 5.18 he suggested 13 principles which the most effective form of contract in modern conditions should include, and in paragraph 5.26 he recommended that the JCT should be restructured, and produce a new family of documents. His enthusiasm for the JCT and all its works was, clearly, somewhat muted.

The JCT's durability has survived this and other criticisms: many of Latham's recommendations were stillborn: the principal tangible result of *Constructing the Team* was the adjudication provisions of the Housing Grants Construction and Regeneration Act 1996, popularly known as The Construction Act. This was intended to provide a speedy method of recovery of monies due under construction contracts. The Housing Act's provisions have caused much litigation, and whether they comply with the Human Rights Act 1998 has been questioned.

Other legislative activity since the last edition includes the CDM Regulations 1994, the Arbitration Act 1996, and the Contracts (Rights of Third Parties) Act 1999. The JCT has reacted to the changes by updating its forms from time to time.

The Courts have continued to produce cases related to the JCT Forms of Contract. Apart from the flood of cases on the Construction Act, of importance to all construction contracts, one could mention: *Beaufort Developments (Northern Ireland) Ltd. v. Gilbert-Ash Northern Ireland Ltd*, overruling *Northern Regional Health Authority v. Derek Crouch Construction Co. Ltd.*, *Oxford University Fixed Assets Ltd. v. Architects Design Partnership*, and the earlier *Crown Estate Commissioners v. John Mowlem & Co. Ltd.* relating to final certificates. The arbitration provisions, including those relating to joinder of sub-contractor's claims have been scrutinised in such cases as *Trafalgar House Construction (Regions) Ltd. v. Railtrack Plc*, and *University of Reading v. Miller Construction Ltd.* Again, changes in the JCT forms have resulted from Court decisions.

More generally, since the last edition, the Official Referees have transformed themselves into Judges of the Technology and Construction Court, and the High Court and County Courts have acquired a new set of rules, the Civil Procedure Rules 1998 (CPR). There is now a specific Practice Direction governing the Technology and Construction Court, and a pre-action protocol which parties are expected to follow before they launch themselves into litigation. These do not directly affect the JCT forms, but they have altered the legal landscape relevant to construction disputes in general.

Despite Sir Michael Latham's strictures on the JCT, the 1998 Form of Contract is very much based on the 1980 edition. It now runs to well over 100 pages, over three times as long as GC/Works/1 (1998). The general view in the industry is not favourable. The perception that it is weighted against the employer has spawned a healthy 'industry' of amendments and additions, including at least one textbook devoted to the subject. The absence of appropriate provisions for collateral warranties, bonds, and insurance provisions to protect the employer have also attracted attention. However the position remains that, as Sir Michael Latham acknowledged, most substantial building work is let under contracts based on JCT 98.

Many who deal with the JCT contracts discover that locating the right point is like searching for a needle in a haystack but whether in drafting contractual provisions, contract administration, or in giving legal advice, Richard Woolley's new book will, like its predecessor, be an invaluable aid to all those professionals and others concerned with the construction industry.

Stephen Bickford-Smith BA (Oxon) FCIArb
Registered Chartered Arbitrator
Co-Editor of *Emden's Construction Law*

Clause No	Clause Title	Signpost
	Part 1	
11	Access to the Works	**Principal applicable clause**
13.1.2.1	Variations/Prov. Sums	Alteration of obligations etc. in respect of access
25.4.12	Extension of time	Failure to give ingress or egress may be a Relevant Event
26.2.6	Loss and Expense	Failure to give ingress or egress may be a ground for claim
28.2.2.4	Determination by Contractor	Failure to give ingress or egress may be a ground for giving notice
	Part 2	
35.6.3	Procedure	On issuing NSC/N the Architect shall confirm any access alterations

ACTIVITY SCHEDULE

Clause No	Clause Title	Signpost
Recital 2	Articles	Contractor to supply Employer with Activity Schedule
	Part 1	
1.3	Definitions etc.	Activity Schedule
30.2.1.1	Certificates/payments	Evaluation of work subject to Retention included in an Activity Schedule
	Appendices	
30.2.1.1	Appendix	Gross valuation

Clause No	Clause Title	Signpost
Art.5	Articles	Disputes may be referred to Adjudication
	Part 1	
41A	Adjudication	**Principal applicable clause**
1.3	Definitions etc.	Adjudication Agreement
1.3	Definitions etc.	Adjudicator
13.4.1.2.A4.3	Contractor's Price Statement	Disputes over Price Statement may be referred to Adjudicator (PS not accepted)
13.4.1.2.A5	Contractor's Price Statement	Disputes over Price Statement may be referred to Adjudicator (PS ignored)
13.4.1.2.A6	Contractor's Price Statement	Procedure where Price Statement disputed but no reference to Adjudicator is made
30.9.4	Certificates/payments	Timing of Arbitration when Adjudicator finds after issue of Final Certificate
	Appendices	
41A.2	Appendix	Appointment of Adjudicator
VAT 3	VAT Agreement	Reference to adjudication of an appeal under VAT rules
VAT 4	VAT Agreement	Reference to adjudication of an appeal under VAT rules
VAT 5	VAT Agreement	Reference to adjudication of an appeal under VAT rules

ARBITRATOR/ARBITRATION

Clause No	Clause Title	Signpost
Art.7	Articles	Disputes to be referred to Arbitration if cl.41B applies
	Part 1	
41B	Arbitration	**Principal applicable clause**
1.3	Definitions etc.	Arbitrator
19.1.2	Assignment	Employer may assign right to bring proceedings in name of the Employer
30.9.1	Certificates/payments	Status of Final Certificate in any proceedings
30.9.2	Certificates/payments	Effect of Arbitration before issue of Final Certificate
30.9.3	Certificates/payments	Effect of Arbitration after issue of Final Certificate
30.9.4	Certificates/payments	Timing of Arbitration when Adjudicator finds after issue of Final Certificate
41A.5.3	Adjudication	Adjudicator does not act as Arbitrator
	Part 2	
36.4.8	Nominated Suppliers	Treatment of disputes

Clause No	Clause Title	Signpost
	Appendices	
41B; 41C	Appendix	Dispute or difference – settlement of disputes – Articles 7A and 7B
41B.1	Appendix	Appointment of Arbitrator
VAT 5	VAT Agreement	Applies to amounts awarded by an Arbitrator

ARCHITECT MAY ...

Clause No	Clause Title	Signpost
	Part 1	
4.3.2.2	Architect's instructions	Give written confirmation of verbal instructions at any time
7	Levels and setting out	Instruct that Contractor errors shall not be amended
8.3	Work, mats. and goods	Issue instructions requiring opening for inspection
8.4.1	Work, mats. and goods	Issue instructions requiring removal of work or materials
8.4.2	Work, mats. and goods	Allow non-compliant work, materials or goods to remain
8.4.3	Work, mats. and goods	Issue instructions to cover 8.4.1 and/or .2 above
8.4.4	Work, mats. and goods	Issue instructions to open up executed work
8.5	Work, mats. and goods	Issue instructions concerning any failure by the Contractor to comply with cl.8.1.3
8.6	Work, mats. and goods	Issue instructions excluding any person from the site
13.2.1	Variations/Prov. Sums	Issue instructions requiring a Variation
13.2.4	Variations/Prov. Sums	Sanction in writing Variations already made
17.3	Practical Compl./Defects	Issue instructions requiring rectification of defects
23.2	Possession, completion	Issue instructions regarding postponement
25.2.3	Extension of time	Require further notices in respect of cls.25.2.2.1 and 2
25.3.2	Extension of time	Fix an earlier Completion Date when work is omitted
25.3.3	Extension of time	Confirm the existing or fix a further new Completion Date within the time stated

Clause No	Clause Title	Signpost
27.2	Determination/Employer	Give notice specifying defaults defined under cl.27.1
	Part 2	
35.5.2	Procedure	Issue instructions to remove objection to an NSC
35.17	Early Final payment	Issue an Interim Certificate to include Sub-Contract Sum
35.24.6.1	Re-nomination	State that the Contractor must obtain a further instruction
35.24.7.2	Re-nomination	Withhold consent to determine NSC employment
	Part 5	
42.5	Performance Specified Work	Require the Contractor to amend a deficient Statement
42.11	Performance Specified Work	Issue instructions requiring Variations

ARCHITECT'S OPINION

Clause No	Clause Title	Signpost
Art.3	Articles	Replacement Architect may not overrule the opinion of his predecessor
	Part 1	
2.1	Contractor's obligations	Approval of quality and standards
5.4.2	Contract documents	Timing issue of further information relative to progress on site and Completion Date
17.1	Practical Compl./Defects	Achievement of Practical Completion
17.4	Practical Compl./Defects	Rectification of defects
18.1.2	Partial possession	Rectification of defects in the Relevant Part
25.3.1	Extension of time	Delay caused by a Relevant Event
25.3.1	Extension of time	If Completion Date is not to be adjusted
25.3.2	Extension of time	Fixing of an earlier Completion Date
25.3.3	Extension of time	Adjustment of Completion Date where the existing date is before Practical Completion
26.1	Loss and expense	Disturbance of regular progress of the Works
26.1.2	Loss and expense	Information concerning disturbance of regular progress
26.4.1	Loss and expense	Disturbance of regular progress of Sub-Contract Works
33.1.3	War Damage	Fixing a later Completion Date
34.3.1	Antiquities	Loss and expense through discovery of antiquities

Clause No	Clause Title	Signpost
	Part 2	
35.16	Practical Completion	Achievement of Practical Completion
35.17.1	Early Final Payment	Remedying of defects
35.24.1	Re-nomination	Default of NSC
36.3.2	Nominated Suppliers	Expense incurred in obtaining materials
36.4.1	Nominated Suppliers	Approval of quality and standards
	Part 4	
41A.5.5.2	Adjudication	Adjudicator may review any opinion
41B.2	Arbitration	Arbitrator may review any opinion
	Part 5	
42.5	Performance Specified Work	Deficiencies in Contractor's Statement

ARCHITECT'S SATISFACTION

Clause No	Clause Title	Signpost
	Part 1	
2.1	Contractor's obligations	Quality of materials and standards of workmanship
8.1.1	Work, materials and goods	Kinds and standards of materials and goods
8.1.2	Work, materials and goods	Standards of workmanship
8.2.2	Work, materials and goods	Architect to express dissatisfaction within a reasonable time
8.4.4	Work, materials and goods	Approval of quality and standards
30.9.1.1	Certificates/payments	Establishing the likelihood or extent of any further similar non-compliance
	Part 2	
35.24.7.2	Re-nomination	Ability of the NSC to continue and carry out the Sub-Contract
36.4.1	Nominated Suppliers	Quality and standards of materials or goods as a matter of Architect's opinion

Clause No	Clause Title	Signpost
Art.3	Articles	*Not* be entitled to disregard or overrule decisions of a predecessor
	Part 1	
2.3.5; 2.4.1	Contractor's obligations	Issue instructions on discrepancies notified to him
4.2	Architect's instructions	Comply with a request for the source of his authority
4.3.1	Architect's instructions	Issue all instructions in writing
5.2	Contract Documents	Issue Contract Documents etc. immediately after execution of the contract
5.3.1.1	Contract Documents	Issue descriptive schedules etc. so soon as possible after execution of the contract
5.4.1	Contract Documents	Ensure information release at time stated in Information Release Schedule
5.4.2	Contract Documents	Issue in good time further details/drawings/instructions which are reasonably necessary
5.7	Contract Documents	*Not* divulge rates or prices in the Contract Bills
5.8	Contract Documents	Issue all certificates to the Employer
5.8	Contract Documents	Send a copy of all certificates to the Contractor
6.1.3	Statutory obligations	Issue instructions on divergence from Contract obligations
6.1.6	Statutory obligations	Issue instructions on divergence from Contractor's Statement
7	Levels and setting out	Determine levels and provide setting-out information
8.2.2	Work, materials and goods	Express any dissatisfaction within a reasonable time

ARCHITECT SHALL ...

Clause No	Clause Title	Signpost
13.3	Variations/Prov.Sums	Issue instructions on expenditure of Provisional Sums
13A.3.2	Variation Instruction	Immediately confirm in writing Employer's acceptance of 13A Quotation
13A.4	Variation Instruction	Issue instructions following Employer's non-acceptance of 13A Quotation
17.1	Practical Compl./Defects	Issue a certificate when Practical Completion is achieved
17.2	Practical Compl./Defects	Specify defects appearing in Defects Liability Period
17.2	Practical Compl/Defects	Deliver schedule of defects within time stated
17.4	Practical Compl./Defects	Issue a certificate when all defects have been made good
18.1	Partial Possession	Issue a written statement identifying relevant part and relevant date
18.1.2	Partial Possession	Issue a certificate when defects have been made good
19.2.2	Assignment/Sub-Contracts	*Not* unreasonably delay or withhold consent to the sub-letting of any portion
22D.1	Insurance: loss of LADs	Notify Contractor upon his entering the Contract whether or not insurance is required
24.1	Damages	Issue a certificate if Completion Date or revised Completion Date is not achieved
25.3.1	Extension of time	Give an extension for delay caused by a Relevant Event
25.3.1	Extension of time	State Relevant Event and the effect of omissions
25.3.1	Extension of time	Fix a new Completion Date within the time stated
25.3.1	Extension of time	Notify Contractor in writing of a decision not to fix a later Completion Date
25.3.3	Extension of time	Confirm existing or fix a further new Completion Date within the time stated

Clause No	Clause Title	Signpost
25.3.5	Extension of time	Notify NSCs of decisions under cl.25.3
25.3.6	Extension of time	*Not* fix a Completion Date that pre-dates the Date for Completion
26.1	Loss and expense	Ascertain/instruct QS to ascertain, loss and expense on disturbance of progress
26.3	Loss and expense	State extensions given under cl.25
26.4.1	Loss and expense	Ascertain/instruct QS to ascertain, loss and expense on Sub-Contract progress
26.4.2	Loss and expense	State in writing any relevant revised periods for completion attributable to NSC
27.4.4	Determination by Employer	Certify loss and expense caused by determination
30.1.1.1	Certificates/payments	Issue Interim Certificates at times as provided in Clause 30
30.1.3	Certificates/payments	*Not* have to issue two Interim Certificates in one month
30.4A.1	Certificates/payments	Prepare or instruct QS to prepare statement of what Retention would have been (Bond)
30.5.2.1	Certificates/payments	Prepare or instruct QS to prepare a Statement of Retention
30.5.2.2	Certificates/payments	Issue Statement of Retention to Employer, Contractor, NSCs
30.6.1.2.1	Certificates/payments	Ascertain or instruct QS to ascertain loss and expense
30.6.1.1	Certificates/payments	Send copy of cl.30.6.1 ascertainments to Contractor with relevant extracts to NSCs
30.7	Certificates/payments	Issue an Interim Certificate for all final NSC accounts
30.8.1	Certificates/payments	Issue the Final Certificate at the required time
30.8.1	Certificates/payments	Notify all NSCs of issue of Final Certificate

ARCHITECT SHALL ...

Clause No	Clause Title	Signpost
34.2	Antiquities	Issue instructions with regard to a reported object
34.3.1	Antiquities	Ascertain or instruct the QS to ascertain loss and expense
34.3.2	Antiquities	If necessary, state extension given in the event of loss and expense
	Part 2	
35.6	Procedure	Issue an instruction on NSC/N, nominating Sub-Contractor
35.6	Procedure	Send Sub-Contractor copy of instruction and Main Contract Appendix
35.9	Procedure	Fix new date for compliance with cl.35.7 or inform Contractor alternatively
35.13.1.1	Payment	Direct the Contractor as to the amount in each Interim Certificate
35.13.1.1	Payment	Compute amounts of interim or final payment in accordance with NSC/C
35.13.1.2	Payment	Inform each NSC of amount included in Interim or Final Certificate
35.13.5.1	Payment	Issue a certificate, with copy to NSC, stating proof of payment not provided
35.14.2	Extension of period	Provisions of NSC/C in response to request for extension of completion period
35.15.1	Extension of period	Issue certificate, - copy to NSC, upon failure of NSC to complete Sub-Contract Works
35.15.2	Extension of period	Issue 35.15.1 certificate not later than 2 months from date of notification by Contractor
35.16	Practical Completion	Issue a certificate with copy to NSC when Practical Completion is achieved
35.16	Practical Completion	Send a written Cl.18 statement to NSC upon Practical Completion of relevant part
35.17	Early Final Payment	Issue an Interim Certificate to include the Sub-Contract Sum

Clause No	Clause Title	Signpost
35.18.1.1	Early Final Payment	Issue an instruction nominating a person to rectify defects
35.24.6.1	Re-nomination	Issue an instruction to give NSC notice of default
35.24.6.3	Re-nomination	Make further nomination as may be necessary
35.24.7.2	Re-nomination	Give consent to determination in cases of insolvency if continuation does not apply
35.24.7.3	Re-nomination	Make further nomination as necessary where consent to determination has been given
35.24.7.4	Re-nomination	Make further nomination as may be necessary where cl.35.24.4 applies
35.24.8.1	Re-nomination	Make further nomination as may be necessary where cl.35.24.3 applies
35.24.8.2	Re-nomination	Make further nomination as may be necessary where cl.35.24.5 applies
35.24.8.10	Re-nomination	Make further nominations within a reasonable time
35.26.1	Determination/Cl.7.1 – 7.5)	Issue information and certificate to enable Contractor to comply with NSC/C cl.7.5.2
35.26.1	Determination/Cl.7.1 – 7.5	Certify values of work, goods and materials not previously certified
35.26.2	Determination/Cl.7.7 – 7.8)	Certify values of work, goods and materials not previously certified
36.2	Nominated Suppliers	Issue instructions for nominating supplier
36.4	Nominated Suppliers	Only nominate a person who will enter into a contract of sale

ARCHITECT SHALL ...

Clause No	Clause Title	Signpost
	Part 5	
42.6	Performance Specified Work	Give Contractor notice of deficiencies which would adversely affect performance
42.14	Performance Specified Work	Give any instructions necessary for integration of Performance Specified Work
42.16	Performance Specified Work	*Not* give time extension under cls.25.3 and 26.1 if Contractor's Statement late

Clause No	Clause Title	Signpost
Articles	Articles of Agreement	**Principal applicable section**
	Part 1	
1.1	Definitions etc.	Reference to Clauses means to Clauses of the Conditions
1.2	Definitions etc.	To be read as a whole with Conditions and Appendix
1.2	Definitions etc.	To be read as subject to qualification
1.3	Definitions etc.	Definitions given in the table are valid unless the Articles otherwise provide
1.3	Definitions etc.	Articles of Agreement
1.3	Definitions etc.	Architect – see Article 3
1.3	Definitions etc.	Conditions – identification of documents annexed to the Articles
1.3	Definitions etc.	Contract Documents – include the Articles
1.3	Definitions etc.	Contract Sum – see Article 2
1.3	Definitions etc.	Contractor – named in the Articles
1.3	Definitions etc.	Employer – named in the Articles
1.3	Definitions etc.	Parties – named in the Articles
1.3	Definitions etc.	Party – named in the Articles
1.3	Definitions etc.	Planning Supervisor – named in Article 6.1

ARTICLES OF AGREEMENT

Clause No	Clause Title	Signpost
1.3	Definitions etc.	Quantity Surveyor – named in Article 4
1.6	Definitions etc	Replacement of Planning Supervisor/Principal Contractor pursuant to Article 6.1 or 6.2
2.2.1	Contractor's obligations	Articles to take precedence over Contract Bills
6A.3	CDM Regulations	Applies from time of replacement of Contractor as Principal Contractor - article 6.2
30.9.1	Certificates/payments	Final Certificate shall have effect in proceedings conducted under Articles 5, 7A or 7B

Clause No	Clause Title	Signpost
	Part 1	
19.1	Assignment/Sub-Contracts	**Principal applicable clause**
6.3	Statutory obligations	Statutory Undertakers pursuing Statutory obligations
27.1.4	Determination by Employer	Failure to comply with cl.19.1.1 or 19.2.2
27.4.2.1	Determination by Employer	Assignment by Contractor to Employer of supply contracts
28.2.1.3	Determination by Contractor	Failure to comply with cl.19.1.1
	Appendices	
19.1.2	Appendix	Assignment by Employer of benefits after Practical Completion
8	Advance Payment Bond	Not assignable without consent of Surety
10	Off-site Materials etc. Bond	Not assignable without consent of Surety
5	Retention Bond	Not assignable without consent of Surety

BASE DATE

Clause No	Clause Title	Signpost
	Part 1	
6.1.7	Statutory obligations	Effect of change in Statutory Requirements of PSW
13.5.4.1	Variations/Prov. Sums	Date of Definition of Prime Cost of Daywork issued by RICS
13.5.4.2	Variations/Prov. Sums	Date of Definition of Prime Cost of Daywork issued by appropriate trade body
15.3	Value added tax	Procedure where supplies become exempt after Base Date
22FC.5	Joint Fire Code	Procedure where Joint Fire Code is altered after Base Date
25.4.9	Extension of time	Effect of exercise by government of statutory power after Base Date as Relevant Event
25.4.10.1	Extension of time	Inability to secure labour as Relevant Event
25.4.10.2	Extension of time	Inability to secure goods and materials as Relevant Event
25.4.15	Extension of time	Delay caused by change in Statutory Requirements on PSW as Relevant Event
	Part 3	
38	Clause 38 provisions etc.	**Principal applicable clauses**
39	Clause 39 provisions etc.	
40.1.1.1	Price adjustment formulae	Contract Sum adjusted in accordance with Formula Rules current at Base Date
40.3	Price adjustment formulae	Market price of articles current at Base Date

Clause No	Clause Title	Signpost
	Part 4	
41B	Arbitration	Reference to Rules means to CIMAR current at Base Date
41B.6	Arbitration	Arbitration to be conducted in accordance with CIMAR current at Base Date
	Appendices	
Fourth recital	Appendix	Statutory tax deduction scheme
1.3	Appendix	Base Date

CERTIFICATES (GENERAL)

Clause No	Clause Title	Signpost
Art. 3	Articles	Replacement Architect may not overrule any certificate etc. of his predecessor
	Part 1	
1.3	Definitions etc.	Certificate of Completion of Making Good Defects (CCMGD.)
1.5	Definitions etc.	Contractor remains responsible for the Works despite any payment certificate issued
1.5	Definitions etc.	Contractor remains responsible for the Works irrespective of any CCMGD issued
17.4	Practical Compl./Defects	Issue of CCMGD
17.4	Practical Compl./Defects	Date of completion of making good defects
18.1	Partial possession	Certificate stating estimated value of Relevant Part
18.1.2	Partial possession	Certificate upon making good defects in Relevant Part
22.3.1	Insurance of the Works	Joint names policy under cls.22A/B/C to cover NSCs to date of CPC of NSC works
22.3.2	Insurance of the Works	Joint names policy under cls.22A/B/C to cover other Sub-Contractors
24.1	Damages	Certificate upon failure to complete by Completion Date
24.1	Damages	Further certificate to be issued if a new Completion Date fixed
24.2.1	Damages	C1.24.1 certificate as precondition to liquidated damages
24.2.3	Damages	Further cl.24.1 certificate does not affect action taken under cl.24.2.1 unless withdrawn
27.6.3.2	Determination by Employer	Statement of accounts may be set out in a certificate issued by the Architect

Clause No	Clause Title	Signpost
28.2.1.1	Determination by Contractor	Failure by the Employer to pay on a certificate
28.2.1.2	Determination by Contractor	Interference with the issue of a certificate
30.1.3	Certificates/payments	Interim Certificate may be issued upon issue of CCMGD.
30.4.1.3	Certificates/payments	Retention release varies if CCMGD, or certificates under cls. 18.1.2 or 35.17 issued
30.8.1	Certificates/payments	Issue of CCMGD is one of the events to precede issue of Final Certificate
30.10	Certificates/payments	Save cl.30.9, no certificate is conclusive evidence of compliance with contract.
31.1	CIS	'Tax certificate' under S.561 of Income and Corporation Taxes Act 1988
	Part 2	
35.13.5.1	Payment	Certificate to the effect that proof of payment not provided
35.13.5.1	Payment	Copy of cl.35.13.5.1 certificate to NSC
35.13.5.2	Payment	Reduction of future payments provided cl.35.13.5.1 certificate issued
35.15.1	Failure to complete	Architect shall certify on failure by NSC to complete
35.15.1	Failure to complete	Duplicate certificate to be sent to NSC
35.15.2	Failure to complete	Timing for issue of cl.35.15.1 certificate
35.16	Practical Completion	Certificate upon achievement of Practical Completion
35.16	Practical Completion	Duplicate of certificate to NSC

CERTIFICATES (GENERAL)

Clause No	Clause Title	Signpost
35.17	Early Final Payment	Timing for Interim Certificate including Sub-Contract Sum
Part 3		
38.4.5	Clause 38 provisions	Applications for Domestic Sub-Contractors to include a Contractor's certificate
39.5.5	Clause 39 provisions	Applications for Domestic Sub-Contractors to include a Contractor's certificate
40.1.3	Price adjustment formulae	Adjustment to be effected in all certificates for payment
40.1.4	Price adjustment formulae	Adjustment of amounts included in previous certificates
Part 4		
41A.5.5.2	Adjudication	Adjudicator may open up, review, revise certificates as if none had been issued
41B.2	Arbitration	Arbitrator may ascertain and award sums that should have been included in certificates
41B.2	Arbitration	Arbitrator may open up, review, revise certificates as if none had been issued
Appendices		
VAT 1.1	VAT Agreement	Timing of final statement of values
VAT 1.2.1	VAT Agreement	VAT added to the Architect's Certificate and remitted within period for payment

Clause No	Clause Title	Signpost
VAT 1.3.1	VAT Agreement	Contractor to issue written final VAT statement after issue of CCMGD
VAT 1.4	VAT Agreement	Contractor is to issue receipts for payment
VAT 2.2	VAT Agreement	Liquidated damages are disregarded
VAT 5	VAT Agreement	Awards by an Arbitrator affecting amounts certified
VAT 8	VAT Agreement	Additional tax arising from determination
4. (iii).c	Retention Bond	Employer's demand includes certificate stating Contractor has failed to complete
6(i)	Retention Bond	Issue of CCMGD is one of the occurrences upon which Surety may be released

CERTIFICATES (INTERIM)

Clause No	Clause Title	Signpost
	Part 1	
30.1 - 30.5	Certificates/payments	**Principal applicable clauses**
1.3	Definitions etc	Interim Certificate
1.3	Definitions etc	Period of Interim Certificates
1.5	Definitions etc.	Contractor responsible for the Works irrespective of any payment certificate issued
3	Contract Sum	Value of authorised adjustments to be included
5.8	Contract Documents	All certificates are to be issued to the Employer
5.8	Contract Documents	Copies of all certificates are to be sent to the Contractor
16.1	Mats. and goods unfixed	Materials included become Employer's property
16.2	Mats. and goods unfixed	Off-site materials included become Employer's property
19.4.2.2	Assignment/Sub-Contracts	Materials included become Employer's property
19.4.2.3	Assignment/Sub-Contracts	Materials are Contractor's property if paid inclusion in an Interim Certificate
22A.4.4	All Risks – by Contractor	Insurance monies paid at Period of Interim Certificate
30.7	Certificates/payments:	Final sums for all NSCs to be certified at least 28 days before issue of Final Certificate
30.8.1	Certificates/payments:	Final Certificate does not prejudice Contractor on sums withheld under cl.30.1.1.4
30.8.1.1	Certificates/payments:	Final Certificate states sum of amounts already stated as due under Interim Certificates
30.10	Certificates/payments	Final Certificate as evidence of compliance with the Contract

Clause No	Clause Title	Signpost
Part 2		
35.13.1	Payment	Procedure on the issue of each Interim Certificate
35.13.3	Payment	Proof of discharge to be provided before Interim Certificate
35.13.5.3.1	Payment	Timing of reduction made under cl.35.13.5.2
35.13.6.2	Payment	Employer may deduct monies for amounts paid to NSC before nomination
35.17	Early Final Payment	Interim Certificate including Sub-Contract Sum
35.24.9	Re-nomination	Include amounts for re-nominated NSC in Interim Certificates
35.26.1	Determination/Cl.7.1 – 7.5)	Interim Certificate to take account of Employer's expenses, loss and damage
35.26.1	Determination/Cl.7.1 – 7.5)	Interim Certificate to certify value of work, goods and material not previously certified
35.26.2	Determination/Cl.7.7)	Interim Certificate to certify value of work, goods and material not previously certified
Part 3		
38.4.7	Clause 38 provisions	Payment only for items not affected by certain events occurring after Completion Date
39.5.7	Clause 39 provisions	Payment only for items not affected by certain events occurring after Completion Date
40.2	Price adjustment formulae	Interim valuations to be made before issue of Interim Certificates
40.6.1	Price adjustment formulae	Adjustment basis if Monthly Bulletins unavailable prior to issue of Final Certificate
40.6.2	Price adjustment formulae	Normal valuation methods reapplied upon resumed publication of Monthly Bulletins

CERTIFICATES (INTERIM)

Clause No	Clause Title	Signpost
40.7.1.1	Price adjustment formulae	Freeze index numbers at Completion Date for work completed after that date
40.7.1.2	Price adjustment formulae	Adjust any errors for calculations not in accordance with cl.40.7.1.1
	Appendices	
30.1.3	Appendix	Period of Interim Certificates
	Off-site Materials Bond	Notice of Demand - agreement to include amounts in Interim Certificates
VAT 1A.2	VAT Agreement	Contractor to give notice of rates by 7 days before date of first Interim Certificate
VAT 1A.3	VAT Agreement	Each Interim Certificate to show an amount calculated for VAT
VAT 1A.3	VAT Agreement	Amounts calculated to be paid within period for payment of Interim Certificates
VAT 1. 1	VAT Agreement	Timing of provisional assessments
VAT 2.3	VAT Agreement	For cl.1A, Employer to pay VAT in Interim Certificate despite deductions under cl.24
VAT 5	VAT Agreement	Awards by an Adjudicator or Arbitrator affecting amounts certified
VAT 8	VAT Agreement	Additional tax arising from determination
1	Retention Bond	Retention will not be deducted from Interim Certificates
2	Retention Bond	Employer to notify Surety of date of next Interim Certificate after issue of CPC

Clause No	Clause Title	Signpost
	Part 1	
30.6 - 30.10	Certificates/payments	**Principal applicable clauses**
1.3	Definitions etc.	Final Certificate
4.3.2.2	Architect's instructions	Confirmation of instructions prior to Final Certificate
24.2.1	Damages	Timing of notification concerning liquidated damages
24.2.1	Damages	Liquidated damages may be deducted
30A	Certificates/payments	Payments subject to cl.31 if Employer is 'contractor' before Final Certificate paid
31.2.2	CIS	Employer as a 'contractor' prior to Final Certificate
	Part 2	
35.13.3	Payment	Proof of discharge to be provided before Final Certificate
35.18.1.1	Early Final payment	Failure to remedy defects before Final Certificate
	Part 3	
38.4.7	Clause 38 provisions	Payment only for items not affected by certain events occurring after Completion Date
39.5.7	Clause 39 provisions	Payment only for items not affected by certain events occurring after Completion Date
40.6.1	Price adjustment formulae	Adjustment basis if Monthly Bulletins unavailable prior to issue of Final Certificate

CERTIFICATES (FINAL)

Clause No	Clause Title	Signpost
40.6.2	Price adjustment formulae	Normal valuation methods reapplied upon resumed publication of Monthly Bulletins
	Appendices	
1A.2	VAT Agreement	Notify rates for use in Final Certificate by 7 days before date of first Interim Certificate
1A.3	VAT Agreement	Final Certificate to show an amount calculated for VAT
VAT 1.1	VAT Agreement	Timing of final provisional assessment
VAT 1.3.1	VAT Agreement	Where cl.1A operates, Cl.1.3 applies only if no VAT shown on Final Certificate
VAT 1.3.2	VAT Agreement	Timing of issue of statement of value of supplies
VAT 2.3	VAT Agreement	For cl.1A, Employer to pay VAT in Interim Certificate despite deductions under cl.24
VAT 4	VAT Agreement	Payment of VAT in Final Certificate, may discharge Employer from further obligation.
1.4.3	EDI	Final Certificate must be in writing and not on electronic data

Clause No	Clause Title	Signpost
	Part 1	
1.5	Definitions etc.	Contractor remains responsible for the Works irrespective of any CPC issued
5.8	Contract documents	All certificates are to be issued to the Employer
5.8	Contract documents	Copies of all certificates are to be sent to the Contractor
17.1	Practical Compl./Defects	Issue of Certificate of Practical Completion (CPC)
17.5	Practical Compl./Defects	Certification on timing of injury causing frost damage
18.1	Partial possession	Employer may take possession of a part of the Works any time prior to issue of CPC
20.3.1	Injury and indemnity	'Property real or personal' in cl.20.2 excludes Works, etc. prior to issue of CPC.
21.2.1.5	Insurance: persons etc.	Joint names policy under cl.21.2.1 to exclude Work and materials subject to CPC
22A.1; B.1	All Risks – Contractor	Insurance maintained until issue of CPC
22C.1	Insurance - existing structures	Insurance maintained until issue of CPC – existing structures
22C.1	Insurance - existing structures	Insurance maintained until issue of CPC – Works
30.1.3	Certificates/payments	Interim Certificates issued at Period for issue until issue of CPC
30.4.1.3	Certificates/payments	Half Retention may be released at Practical Completion
	Appendices	
17.2	Appendix	Defects Liability Period – if none stated 6 months from day named in CPC

CLERK OF WORKS

Clause No	Clause Title	Signpost
	Part 1	
12	Clerk of Works	**Principal applicable clause**
1.5	Definitions etc.	Contractor responsibility whether or not Clerk of Works appointed
1.5	Definitions etc.	Contractor responsibility whether or not Clerk of Works inspects work in preparation
10	Person-in-charge	Directions by Clerk of Works to person-in-charge
34.1.3	Antiquities	Notification of discovery

Clause No	Clause Title	Signpost
	Part 1	
25	Extension of time	**Principal applicable clause**
1.3	Definitions etc.	Completion Date
5.4.2	Contract documents	Timing for provision of further drawings or details
13A.2.2	Variation instruction	Revised Completion Date may be earlier than Date for Completion
13A.2.2	Variation instruction	13A Quotation to include for time adjustments relative to Completion Dates
13A.3.2.3	Variation instruction	Acceptance of 13A Quotation to include revised Completion Date
13A.3.2.3	Variation instruction	Revised Completion Date may be earlier than Date for Completion
22FC.3.1	Joint Fire Code	Definition of and obligations towards Remedial Measures Completion Date
23.1.1	Possession, completion	The Contractor is to complete on or before Completion Date
24.1	Damages	Certificate issued if Completion Date or revised Completion Date is not achieved
24.2.1.1	Damages	Period for which liquidated damages are calculated
24.2.1.2	Damages	Deduction of liquidated damages
24.2.2	Damages	Repayment of liquidated damages
	PART 3	
38.4.7	Clause 38 provisions	Circumstances in which no adjustments are made after Completion Date
38.4.8.2	Clause 38 provisions	Revised Completion Date in writing as a condition for cl.38.4.7 to apply

COMPLETION DATE

Clause No	Clause Title	Signpost
39.5.7	Clause 39 provisions	Circumstances in which no adjustments are made after Completion Date
39.5.8	Clause 39 provisions	Revised Completion Date in writing as a condition for cl.39.5.7 to apply
40.7.1.1	Price adjustment formulae	Rules for adjustment if Contractor fails to complete by Completion Date
40.7.1.2	Price adjustment formulae	Correction to be made if adjustment wrongly made under cl.40.7.1.1

Appendices

4(iii)(c)	Retention Bond	Amount demanded may include liquidated damages

Clause No	Clause Title	Signpost
Art. 2	Articles	Contract Sum adjustable in manner specified in the Conditions
Art. 3	Articles	Identification of 'Architect' as term is used in the Conditions
Art. 4	Articles	Identification of 'Quantity Surveyor' as term is used in the Conditions
Art. 6.1	Articles	Identification of 'Planning Supervisor' as term is used in the Conditions
Art. 6.2	Articles	Identification of 'Principal Contractor' as term is used in the Conditions
Art. 5.2.2	Articles	Whether instructions or certificates accord with Conditions
	Part 1	
1.1	Definitions etc	Reference to Clauses means to Clauses of the Conditions
1.2	Definitions etc	To be read as a whole with Articles and Appendix
1.2	Definitions etc	To be read as subject to qualification
1.3	Definitions etc	Definitions in cl.1.3 apply throughout the Conditions unless otherwise stated
1.3	Definitions etc	Appendix refers to Appendix to Conditions
1.3	Definitions etc	Articles are those to which the Conditions are annexed
1.3	Definitions etc	Conditions
1.5	Definitions etc	Contractor is responsible for carrying out Works in accordance with the Conditions
1.9	Definitions etc	Employer's Representative to carry out Employer's functions according to Conditions

CONDITIONS OF CONTRACT

Clause No	Clause Title	Signpost
2.1	Contractor's obligations	Carrying out the Works is subject to the Conditions
2.2.1	Contractor's obligations	Conditions take precedence over Contract Bills
2.3.3	Contractor's obligations	Discrepancies between Conditions and other documents
3	Contract Sum	Effect of provision for adjustment of Contract Sum
4.1.1	Architect's instructions	Compliance with instructions empowered by the Conditions
4.2	Architect's instructions	The Architect to specify authority for an instruction
4.2	Architect's instructions	Compliance with an instruction after the Condition specified
5.4.2	Contract documents	Drawings and details necessary for work to be done
5.8	Contract documents	Certificates issued under the Conditions are issued to the Employer
6.1.1	Statutory obligations	Requirements referred to as 'the Statutory Requirements'
8.4.4	Work, materials and goods	Instructions for opening up work to accord with Code of Practice appended
12	Clerk of Works	Directions must be for matters empowered by Conditions
13.1	Variations/Prov. Sums	Definition of the term 'Variation'
13.4.1.1	Variations/Prov. Sums	Instructions issued in regard to expenditure of Provisional Sums
13.5.7	Variation/Prov. Sums	Loss and expense reimbursable under any other Condition
13A.3.2	Variation instruction	Definition of 'confirmed acceptance' applicable within the Conditions
14.2	Contract Sum	Adjustment to be only in accordance with the Conditions

Clause No	Clause Title	Signpost
15.2	Value added tax	'Contract Sum' means exclusive of VAT
19.4.2.1	Assignment/Sub-Contracts	A sub-let contract is subject to cl.16.2 of Conditions in regard to unfixed materials etc.
19.4.2.1	Assignment/Sub-Contracts	Conditions referred to as Main Contract Conditions in cls.19.4.2.2.to .4
19.4.2.2	Assignment/Sub-Contracts	Property in unfixed materials etc. part of a sub-let contract and are paid by Employer
19.4.2.3	Assignment/Sub-Contracts	Property in unfixed materials etc. part of a sub-let contract and are paid by Contractor
19.4.2.4	Assignment/Sub-Contracts	Cls. 19.4.2.1 to 3 without prejudice to transfers under cl.30.3 of Conditions
19.5.1	Assignment/Sub-Contracts	Provisions relating to NSCs are set out in Part 2 of the Conditions
19.5.1	Assignment/Sub-Contracts	Contractor is responsible for Works including NSC work unless otherwise stated
22.2	Insurance of the Works	Insurance definitions have meanings given where relevant throughout the Conditions
26.4.2	Loss and expense	Assessment of loss and expense related to work by NSCs
27.2.1.5	Determination by Employer	Failure pursuant to Conditions to comply with CDM Regulations may be a ground
27.7.1.1	Determination by Employer	If Employer does not proceed, Account Statement valued to accord with Conditions
28.2.1.4	Determination by Contractor	Failure pursuant to Conditions to comply with CDM Regulations may be a ground
28.4.3.1	Determination by Contractor	Account Statement to be valued in accordance with Conditions
28A.5.1	Determination by either	Account Statement to be valued in accordance with Conditions
29.1	Works by Employer	Execution of work not forming part of this Contract
30.1.1.1	Certificates/payments	Interest in the event of failure to pay under the Conditions

CONDITIONS OF CONTRACT

Clause No	Clause Title	Signpost
30.1.4	Certificates/payments	Suspension of performance in the event of non-payment by the Employer
30.2	Certificates/payments	Ascertainment of amounts due and Retention held in accordance with the Conditions
30.4.1.1	Certificates/payments	Reference to definition of Retention
30.4A.2	Certificates/payments	Retention Bond as set out in Annex 3 to the Conditions
30.9.1.1	Certificates/payments	Final Certificate as conclusive evidence of quality of work
31.13	CIS	C1.31 prevails over all other Clauses
	Part 2	
35.20	Position of Employer	Nothing apart from NSC/W renders Employer liable to NSC
36.1.1/2	Nominated Suppliers	Definition of the term 'Nominated Supplier'
	Part 3	
38.4.8.1	Clause 38 provisions	Clause 25 must be unamended and form part of the Conditions for cl.38.4.7 to apply
39.5.8.1	Clause 39 provisions	Clause 25 must be unamended and form part of the Conditions for cl.39.5.7 to apply
40.1.3	Price adjustment formulae	Cl.40 applies in all certificates for payment issued under the conditions
40.7.2.1	Price adjustment formulae	Clause 25 must be unamended and form part of the Conditions for cl.40.7.1 to apply

Clause No	Clause Title	Signpost
	Part 5	
42.1.4; 42.2	Performance Specified Work	Work subject to the Conditions
42.2	Performance Specified Work	Documents referred to under the Conditions as 'Contractor's Statement'
42.5	Performance Specified Work	Contractor responsible for deficiency in Statement
42.6	Performance Specified Work	Contractor responsible for Performance Specified Work
	Appendices	
	VAT Agreement	General references to the Conditions
VAT 1A.1	VAT Agreement	Reference to cl.15.2 of Conditions
VAT 1A.3	VAT Agreement	Reference to cl.15.2 of Conditions
VAT 1.2.1	VAT Agreement	Reference to cl.30.1.1.1 of Conditions – period for payment
VAT 1.3.1	VAT Agreement	Reference to cl.17.4 of Conditions – issue of CCMGD
VAT 1.3.2	VAT Agreement	Reference to cl.30.8 of Conditions – issue of Final Certificate
VAT 2.1	VAT Agreement	Reference to cl.24 of Conditions – deduction of liquidated damages
VAT 2.2	VAT Agreement	Ascertaining values to be included in final statement issued under the Conditions
VAT 2.2	VAT Agreement	Liquidated damages to be disregarded in ascertaining values for final statement
VAT 2.3	VAT Agreement	Reference to cl.24 – deduction of liquidated damages when cl.1A operates

CONDITIONS OF CONTRACT

Clause No	Clause Title	Signpost
VAT 3.2	VAT Agreement	Full amount of tax must be paid before appeal from a decision of the Commissioners
VAT 7	VAT Agreement	Obligations for further payment if no VAT receipt given
VAT 8	VAT Agreement	Reference to cl.27.4 – tax adjustment in event of determination by Employer
	EDI	Reference to cl.1.11 of Conditions
1.4.5	EDI	Any amendment to the Conditions must be in writing

Clause No	Clause Title	Signpost
Recital 5	Articles	Extent of application is in Appendix
Art. 6.1	Articles	Appointment of Planning Supervisor
Art. 6.2	Articles	Appointment of Principal Contractor
	Part 1	
6A	CDM Regulations	**Principal applicable clause**
1.3	Definitions etc.	CDM Regulations
1.3	Definitions etc.	Health and Safety Plan
1.3	Definitions etc.	Planning Supervisor
1.3	Definitions etc.	Principal Contractor
27.2.1.5	Determination by Employer	Failure to comply with CDM Regulations as a ground
	Appendices	
Recital 5	Appendix	CDM Regulations

CONTRACT BILLS

Clause No	Clause Title	Signpost
Recital 2	Articles	Contractor has supplied Employer with fully priced copy
Recital 3	Articles	Contract Bills signed by the parties
	Part 1	
1.3	Definitions etc	Activity Schedule – Approximate quantities included in Contract Bills
1.3	Definitions etc	Approximate Quantity - Approximate quantities included in Contract Bills
1.3	Definitions etc	Contract Bills
1.3	Definitions etc	Contract Documents – include Contract Bills
2.2.1	Contractor's obligations	Articles, Conditions or Appendix are not overridden
2.2.2.1	Contractor's obligations	Method of Measurement - SMM7
2.2.2.2	Contractor's obligations	Correction of any departure from Method of Measurement
2.2.2.2	Contractor's obligations	Correction of error
2.3.2	Contractor's obligations	Discrepancy between Contract Bills and other documents
5.1	Contract Documents	Custody of Contract Bills
5.2.3	Contract Documents	Unpriced copies to the Contractor
5.5	Contract Documents	A copy of unpriced bills on site
5.7	Contract Documents	Rates or prices are not to be divulged

Clause No	Clause Title	Signpost
5.9	Contract Documents	Performance Specified Work as may be specified in the Contract Bills
6.2.3	Statutory obligations	Fees and charges to be added to the Contract Sum unless priced in Contract Bills
8.1.1	Materials, goods etc	Materials and goods are to be as described
8.1.2	Materials, goods etc	Standards to be as may be described in Contract Bills
8.3	Materials, goods etc	Cost of inspection of compliant work added to Contract Sum unless already priced
9.1	Royalties etc.	Royalties for all items described in Contract Bills are deemed included
13.1.2	Variations/Prov. Sums	Alteration of work referred to in the Contract Bills
13.1.2	Variations/Prov. Sums	Alteration of obligations imposed in the Contract Bills
13.1.3	Variations/Prov. Sums	Nomination for execution of measured work is not a Variation
13.3.1	Variations/Prov. Sums	Instructions regarding Provisional Sums in the Contract Bills
13.4.1	Variations/Prov. Sums	Valuation of work arising from Provisional Sum expenditure
13.4.1	Variations/Prov. Sums	Valuation of work arising from remeasurement of Approximate Quantities
13.5.1	Variations/Prov. Sums	Valuation of work including Approximate Quantities in Contract Bills
13.5.1.1	Variations/Prov. Sums	Valuation of work similar to that in the Contract Bills executed under other conditions
13.5.1.2	Variations/Prov. Sums	Valuation of work similar to that in the Contract Bills executed under other conditions
13.5.1.3	Variations/Prov. Sums	Valuation of work dissimilar to that in the Contract Bills
13.5.2	Variations/Prov. Sums	Omission of work in the Contract Bills

CONTRACT BILLS

Clause No	Clause Title	Signpost
13.5.2	Variations/Prov. Sums	Omission of work in the Contract Bills
13.5.3.1	Variations/Prov. Sums	Rules for measurement of Variations are the same as for the Contract Bills
13.5.3.2	Variations/Prov. Sums	Percentage or lump sum adjustments
13.5.4.1	Variations/Prov. Sums	Percentage additions in the Contract Bills for Daywork – general work
13.5.4.2	Variations/Prov. Sums	Percentage additions in the Contract Bills for Daywork – specialist work
13.5.5	Variations/Prov. Sums	Valuation of other work where a Variation changes the circumstances for execution
13.5.6.2	Variations/Prov. Sums	Use of rates in Contract Bills to value varied Performance Specified Work
13.5.6.3	Variations/Prov. Sums	Valuation of omitted Performance Specified Work
13.5.6.6	Variations/Prov. Sums	Valuation of other work where a Variation changes the circumstances for execution
13A.2.1	Variation instruction	13A Quotation to employ rates in Contract Bills where appropriate
14.1	Contract Sum	Quality and quantity of work in the Contract Sum
19.3.1	Assignment/Sub Contracts	Measured work carried out by named firms
19.3.2.1	Assignment/Sub Contracts	Lists of names to be attached to the Contract Bills
25.4.12	Extension of time	Failure to give access in accordance with the Contract Bills – Relevant Event
25.4.14	Extension of time	Approximate Quantities not a reasonably accurate forecast – Relevant Event
26.2.3	Loss and expense	Discrepancy between Contract Bills, Contract Drawings and Numbered Documents
26.2.6	Loss and expense	Failure to give access in accordance with the Contract Bills – matters in cl.26.1

Clause No	Clause Title	Signpost
26.2.8	Loss and expense	Approximate Quantities not a reasonably accurate forecast – matters in cl.26.1
29.1	Works by Employer	Execution of work not forming part of this Contract
29.2	Works by Employer	Procedure where information defined in cl.29.1 is not provided
30.2.1.5	Certificates/payments	Profit on NSC work calculated at rates in Contract Bills
30.6.2.1	Certificates/payments	Omission of profit on Prime Cost Sums etc
30.6.2.2	Certificates/payments	Omission of Provisional Sums and provisional work
30.6.2.3	Certificates/payments	Omission of other work affected by a Variation
30.6.2.9	Certificates/payments	Addition of profit on NSC work etc
30.6.2.12	Certificates/payments	Addition of the value of work substituting Provisional Sums
30.6.2.12	Certificates/payments	Addition of the value of work substituting Approximate Quantities
30.9.1	Certificates/payments	Final Certificate as conclusive evidence of quality of work
	Part 2	
35.1.1	General	Nomination of Sub-Contractors where rights reserved
35.1.2	General	Nomination arising from expenditure of a Provisional Sum
35.1.3.1	General	Variations of work additional to that in Contract Bills
35.1.3.2	General	Instructions for further NSC work or goods of similar type
35.1.4	General	Provisions of SMM A51 apply to adjustment of Prime Cost Sums in Contract Bills

CONTRACT BILLS

Clause No	Clause Title	Signpost
35.2.1	General	Contractor's tender for work reserved for NSC
35.2.1	General	Instruction issued under cl.13.3 deemed to have been for work in Bills
35.2.2	General	References to Bills within accepted tender from Contractor
36.1.1.1	Nominated Suppliers	Prime Cost Sum included for supply of materials and goods
36.1.1.1	Nominated Suppliers	Naming of supplier in Contract Bills
36.1.1.2	Nominated Suppliers	Provisional Sum in Contract Bills
36.1.1.3	Nominated Suppliers	Provisional Sum in Contract Bills
36.1.2	Nominated Suppliers	Materials or goods to be fixed by the Contractor
36.1.2	Nominated Suppliers	Prime Cost Sums for materials and goods
36.1.2	Nominated Suppliers	Named supplier in the Contract Bills
36.2	Nominated Suppliers	Instructions to be issued on expenditure of Prime Cost Sums

Part 3

38.1.1	Fluctuations – taxes etc.	Basis of prices in Contract Bills – contributions etc payable by Contractor
38.1.5	Fluctuations – taxes etc.	Basis of prices in Contract Bills – contributions etc payable and receivable
39.1.5	Fluctuations – cost, taxes etc.	Basis of prices in Contract Bills – transport charges attached to Contract Bills
39.2.1	Fluctuations – cost, taxes etc.	Basis of prices in Contract Bills – contributions etc payable by Contractor
39.2.4	Fluctuations – cost, taxes etc.	Basis of prices in Contract Bills – contributions etc payable and receivable

Clause No	Clause Title	Signpost
39.3.1	Fluctuations – cost, taxes etc.	Basis of prices in Contract Bills – market price of materials
39.3.1	Fluctuations – cost, taxes etc.	Basis of prices in Contract Bills – market price of fuels
39.3.1	Fluctuations – cost, taxes etc.	Basis of prices in Contract Bills – list attached to Contract Bills
39.7.2	Clause 39 provisions	Definitions of materials and goods
40.3	Price adjustment formulae	Contractor to attach a list to the Contract Bills of articles to which Formula Rules apply
40.5.1	Price adjustment formulae	Part of Formula Rules applicable is to be stated in the Contract Bills

Part 5

42.1.4	Performance Specified Work	Performance required to be stated in the Contract Bills
42.4	Performance Specified Work	Contractor's Statement to be provided by date given in Contract Bills
42.7	Performance Specified Work	Definition of the Provisional Sum for PSW
42.8	Performance Specified Work	Instructions for PSW limited to expenditure of specific Provisional Sums
42.9	Performance Specified Work	PSW in Contract Bills is not a departure from their method of preparation
42.10	Performance Specified Work	Correction of errors in Contract Bills
42.13	Performance Specified Work	Contractor to provide analysis where Contract Bills do not.

Appendices

42.1.1	Appendix	Identify each portion of Performance Specified Work

CONTRACT SUM

Clause No	Clause Title	Signpost
Art. 2	Articles	Identification of Contract Sum
	Part 1	
3; 14	Contract Sum/Adjustments	**Principal applicable clauses**
1.3	Definitions etc.	Activity Schedule - Sum of activities is the Contract Sum
1.3	Definitions etc.	Contract Sum
6.2	Statutory obligations	Fees and charges – defining whether or not included in Contract Sum
7	Levels and setting out	Adjustment in the event of error
8.3	Materials, goods etc.	Increased for opening up Works unless discovered faulty
8.4.2	Materials, goods etc.	Deduction made if not in accordance with contract but allowed to remain
8.4.3	Materials, goods etc.	Variation not in accordance with contract but remains – no addition to Contract Sum
8.4.4	Materials, goods etc.	Opening for inspection not in accordance with contract – no addition to Contract Sum
8.5	Materials, goods etc.	No addition for extra work arising from work not in accordance with 8.1.3
9.1	Royalties etc	Deemed included
9.2	Royalties etc.	Increased for charges which arise from instructions
13.4.1.2.A3	Contractor's Price Statement	Contract Sum adjusted to accord with effect of accepted Price Statement
13.7	Variations/ Prov. Sums	Addition or deduction gives effect to a Valuation

Clause No	Clause Title	Signpost
13A.2.1	Variation instruction	13A Quotation includes the value of the adjustment to the Contract Sum
13A.3.2.2	Variation instruction	Confirmation of acceptance of a 13A Quotation must include the adjustment
13A.5	Variation instruction	Cost of preparing a 13A Quotation may be added to the Contract Sum
15.2	Value added tax	Exclusive of VAT
17.2; 17.3	Practical Compl./Defects	Deductions when defects are not made good
18.1.4	Partial possession	Proportionate adjustments when works remain incomplete beyond relevant date
21.2.3	Insurance: persons etc	Cost of insurances to be added
22A.5.4.1	All Risks – Contractor	Contract Sum adjusted where terrorism insurance premium alters
22B.2	All Risks – Employer	Amount of premium paid by Contractor may be added
22C.3	Insurance - existing structures	Amount of premium paid by Contractor may be added
22D.4	Insurance: loss of LADs	Amount of premium paid by Contractor may be added
22FC.5	Joint Fire Code	Extra cost of amended Joint Fire Code may be added
26.5	Loss and expense	Loss and expense to be added
30.6.1.1	Certificates/payments	Contractor to provide all documents necessary for adjustment of Contract Sum
30.6.1.2.2	Certificates/payments	Quantity Surveyor to provide a statement of adjustments to Contract Sum
30.6.2	Certificates/payments	Adjustments to be made to Contract Sum
30.6.2.14	Certificates/payments	Any other amounts to be added to the Contract Sum

CONTRACT SUM

Clause No	Clause Title	Signpost
30.8.2	Certificates/payments	Adjusted Contract Sum to be stated in Final Certificate
30.9.1.2	Certificates/payments	Final Certificate evidence that all adjustments made
34.3.3	Antiquities	Amounts ascertained shall be added
Part 2		
35.24.9	Re-nomination	Amounts payable resulting from further nomination are added
36.3.2	Nominated Suppliers	Expenses in obtaining materials from Nominated Suppliers
Part 3		
38.1	Fluctuations – taxes etc.	Contract Sum is deemed to have been calculated in the manner set out in the clause
38.1.1	Fluctuations – taxes etc.	Basis of Contract Sum - as an employer
38.2	Fluctuations – taxes etc.	Contract Sum is deemed to have been calculated in the manner set out in the clause
38.2.1	Fluctuations – taxes etc.	Basis of Contract Sum - as a purchaser of goods and materials and services
38.4.4.1	Clause 38 provisions	Amounts which become payable are added to or deducted from the Contract Sum
38.4.6	Clause 38 provisions	Adjustments do not alter the level of profit of the Contractor
39.1	Fluctuations – cost, taxes etc.	Contract Sum is deemed to have been calculated in the manner set out in the clause
39.2	Fluctuations – cost, taxes etc.	Contract Sum is deemed to have been calculated in the manner set out in the clause

CONTRACTOR MAY …

Clause No	Clause Title	Signpost
Art. 5	Articles	Refer a dispute to Adjudication
	Part 1	
4.2	Architect's instructions	Request Architect to specify authority for instruction
5.4.1	Contract documents	Agree to vary the time by which information may be released
13.4.1.2.A1	Contractor's Price Statement	Submit 'Price Statement' to the Quantity Surveyor
13.4.1.2.A5	Contractor's Price Statement	Refer 'Price Statement' to Adjudicator if no Para. A2 notification given
13.6	Variations/Prov. Sums	Be present when QS takes measurements
13.6	Variations/Prov. Sums	Require notes and measurements to be taken
13A.6	Variation instruction	Not use 13A Quotation for any purpose if no confirmed acceptance issued
13A.7	Variation instruction	Agree to vary time scales stated in cls.13A.1.1 and .2
19.3.2.2	Assignment/Sub-Contracts	Sub-let 'named firm' work when fewer than three available
22B.2	All Risks –Employer	Take out Joint Names Policy if Employer defaults
22C.3	Insurance - existing structures	Take out Joint Names Policy if Employer defaults under cls.22C.1 or 22C.2
22C.4.3.1	Insurance - existing structures	Determine employment of Contractor following loss or damage
22C.4.3.1	Insurance - existing structures	Invoke relevant procedures to decide whether determination was just and equitable
26.1	Loss and expense	Give quantification of direct loss and expense

Clause No	Clause Title	Signpost
27.5.3	Determination by Employer	Make an interim arrangement for work to be carried out
27.7.2	Determination by Employer	Require Employer to state intentions if he has not operated cl.27.6.1 within 6 months
28.2	Determination by Contractor	Give notice specifying defaults
28.2.2	Determination by Contractor	Give notice specifying events causing work to be suspended
28.2.3	Determination by Contractor	Give notice determining his employment on the expiry of given times
28.2.4	Determination by Contractor	Give notice determining his employment on the repetition of certain events
28.3.3	Determination by Contractor	Give notice determining his employment if Employer becomes bankrupt or similar
28A.1.1	Determination by either	Give notice determining his employment if works suspended for no-fault reasons
30.1.2.2	Certificates/payments	Submit an application for payment
30.1.4	Certificates/payments	Suspend operations, after giving proper notices, upon Employer failing to pay in full
30.9.4	Certificates/payments	Commence arbitration or legal proceedings within 28 days of Adjudicator's decision
	Part 2	
35.13.4	Payment	Require evidence from NSC of payment
35.25	Determine	Be entitled to a right to determine employment of NSC but without consent

CONTRACTOR MAY ...

Clause No	Clause Title	Signpost
	Part 3	
38.4.3	Clause 38 provisions	Agree the net amount payable or allowable
39.5.3	Clause 39 provisions	Agree the net amount payable or allowable
40.5	Price adjustment formulae	Agree any alteration to methods and procedures for ascertaining the adjustment
	Part 4	
41A	Adjudication	Several references to matters at the discretion of both parties
41B	Arbitration	Several references to matters at the discretion of both parties
	Appendices	
VAT 1A.4	VAT Agreement	Give written notice that cl.1A shall no longer apply
VAT 1.2.2	VAT Agreement	Treat amounts received as inclusive of tax if assessment confirmed
VAT 1.3.2	VAT Agreement	Issue statement of values at the stated time

Clause No	Clause Title	Signpost
Art. 1	Articles	Carry out and complete Works shown in the Documents
	Part 1	
2.1	Contractor's obligations	Carry out and complete Works in accordance with Documents
2.3; 2.4.1	Contractor's obligations	Give notice of discrepancies between Documents
2.4.2	Contractor's obligations	Correct discrepancies in Contractor's Statement
4.1.1	Architect's instructions	Forthwith comply with all instructions unless objectionable
4.3.2	Architect's instructions	Confirm within 7 days instructions not issued in writing
4.3.2.1	Architect's instructions	*Not* be obliged to confirm instructions if Architect confirms in writing within 7 days
5.3.1.2	Contract Documents	Provide Architect with copies of master programme
5.4.2	Contract Documents	Advise the Architect of timing for issue of further information
5.5	Contract Documents	Keep on site a copy of relevant Documents
5.6	Contract Documents	Return all Documents issued if requested
5.7	Contract Documents	*Not* use Documents for any other purpose
5.9	Contract Documents	Provide Employer with all necessary information, including maintenance and operation
6.1.1	Statutory obligations	Comply with all Statutory Requirements
6.1.2	Statutory obligations	Give notice of divergence with Contract requirements

CONTRACTOR SHALL …

Clause No	Clause Title	Signpost
6.1.4.1	Statutory obligations	Carry out emergency work
6.1.4.2	Statutory obligations	Notify Architect of emergency work
6.1.5	Statutory obligations	*Not* be liable for non-compliance of Works provided he has complied with cl.6.1.2
6.1.7	Statutory obligations	Give notice of divergence with Contractor's Statement
6.2	Statutory obligations	Pay and indemnify Employer against liability
6A.2	CDM Regulations	Comply with duties of a Principal Contractor
6A.2	CDM Regulations	Ensure that Health and Safety Plan has features required by regulation 15(4)
6A.2	CDM Regulations	Notify Employer of changes to Health and Safety Plan
6A.3	CDM Regulations	Comply with all reasonable requirements of Principal Contractor
6A.4	CDM Regulations	Provide information for health and safety file
6A.4	CDM Regulations	Ensure any Sub-Contractor provides information for health and safety file
7	Levels and setting out	Be responsible for and amend his own errors
8.1.4	Work, materials and goods	*Not* substitute materials or goods without consent of the Architect
8.2.1	Work, materials and goods	Provide Architect with vouchers proving quality
8.4.2; 8.4.3;	Work, materials and goods	Consult immediately with relevant NSC over quality questions raised by the Architect
8.5	Work, materials and goods	Consult immediately with relevant NSC over any failure to comply with cl.8.1.3
9.1	Royalties etc	Indemnify Employer against royalty etc. claims

Clause No	Clause Title	Signpost
9.2	Royalties etc	*Not* be liable for any infringement in complying with Architect's instructions
10	Person-in-charge	Keep on site a competent person-in-charge
11	Access to the Works	Secure for Architect right of access to Sub-Contractor workshops
12	Clerk of Works	Afford every reasonable facility of performance of duties
13.4.1.2.A4.2	Contractor's Price Statement	State within 14 days whether he accepts amended Price Statement
13A.1.1	Variation instruction	Request further information if he considers it necessary for a 13A Quotation
13A.1.2	Variation instruction	Submit 13A Quotation to the Quantity Surveyor
13A.1.2	Variation instruction	Include any relevant 3.3A Quotations for NSC work
16.1	Mats. and goods unfixed	Remain responsible for those included in Certificate
16.2	Mats. and goods unfixed	*Not* remove materials from storage except for use on site
16.2	Mats. and goods unfixed	Be responsible for those off-site included in Certificate
17.2	Practical Compl./Defects	Make good defects in schedule at his own cost
17.3	Practical Compl./Defects	Comply with instructions for making good defects
17.5	Practical Compl./Defects	*Not* have to repair certain frost damage
18.1.4	Partial Possession	Reduce value of Clause 22A insurance
19.1.1	Assignment/Sub-Contracts	*Not* assign the contract without written consent
19.2.2	Assignment/Sub-Contracts	*Not* sub-let any part without written consent

CONTRACTOR SHALL …

Clause No	Clause Title	Signpost
19.2.2	Assignment/Sub-Contracts	Remain responsible for all sub-let works
19.3.2.1	Assignment/Sub-Contracts	Be entitled, with consent, to add names to lists of named specialist firms
19.3.2.2	Assignment/Sub-Contracts	Add, by agreement, names to lists of specialist firms if less that three persons named
19.5.1	Assignment/Sub-Contracts	Remain responsible for all works let to NSCs
20.1	Injury and indemnity	Be liable for and indemnify the Employer against death or injury claims
20.2	Injury and indemnity	Be liable for and indemnify the Employer against property damage claims
21.1.1.1	Insurance: persons etc.	Take out and maintain the required insurances
21.1.2	Insurance: persons etc.	Produce evidence of cl.21.1 insurances
21.2.1	Insurance: persons etc.	If instructed, take joint names insurance for damage by non-attributable collapse etc.
21.2.2	Insurance: persons etc.	Deposit with the Employer policies of cl.21.2 insurance
21.3	Insurance: persons etc	*Not* be liable to indemnify or insure against Excepted Risks
22.3.1	Insurance of the Works	Where cl.22A applies, ensure Joint Names Policy includes items specific to NSCs
22A.1	All Risks –Contractor	Take out and maintain All Risks insurance
22A.2	All Risks –Contractor	Send policies of cl.22A.1 insurance to Architect for deposit with the Employer
22A.3.1	All Risks –Contractor	Be discharged from obligation if he has evidence of adequate insurance already held
22A.4.1	All Risks –Contractor	Give notice of loss or damage caused by risks covered by policy
22A.4.3	All Risks –Contractor	Restore damage and proceed upon completion of inspection by insurers

Clause No	Clause Title	Signpost
22A.4.5	All Risks –Contractor	*Not* be entitled to any monies other than those received under the insurance
22A.5.1	All Risks –Contractor	Notify Employer immediately if insurers withdraw terrorism cover
22A.5.3	All Risks –Contractor	Restore damage and proceed
22B.3.1	All Risks – Employer	Give notice of loss or damage caused by risks covered by policy
22B.3.3	All Risks – Employer	Restore damage and proceed upon completion of inspection by insurers
22B.3.4	All Risks – Employer	Authorise insurers to pay insurance monies to the Employer
22B.4.1	All Risks – Employer	Notify Employer immediately if insurers withdraw terrorism cover
22B.4.3	All Risks – Employer	Restore damage and proceed
22C.1	Insurance - existing structures	Authorise insurers to pay insurance monies to the Employer
22C.4; .4.2	Insurance - existing structures	Give notice of loss or damage caused by risks covered by policy
22C.4.4.1	Insurance - existing structures	Restore damage and proceed upon completion of inspection by insurers
22C.5.1	Insurance - existing structures	Restore damage and proceed
22D.1	Insurance for loss of LADs	If requested, take out and maintain insurance
22D.1	Insurance for loss of LADs	Send quotation to the Architect
22FC.2.2	Joint Fire Code	Comply with the Joint Fire Code
22FC.3.1	Joint Fire Code	Ensure that remedial measures are in accordance with the Architect's instructions
22FC.4	Joint Fire Code	Indemnify the Employer for the consequences of a breach of the Joint Fire Code

CONTRACTOR SHALL ...

Clause No	Clause Title	Signpost
23.1.1	Possession, completion	Begin Works on date of possession and proceed diligently
23.1.1	Possession, completion	Complete Works on or before Completion Date
23.3.1	Possession, completion	Retain possession of site and Works until date of issue of CPC
23.3.3	Possession, completion	Notify Employer of any extra insurance premiums for his advance use of site or Works
23.3.3	Possession, completion	Provide Employer with additional premium receipts if requested
25.2.1.1	Extension of time	Give notice of cause of delay and identify Relevant Event
25.2.1.2	Extension of time	Send copy of cl.25.2.1.1 notice to NSC if appropriate
25.2.2	Extension of time	Give particulars of effects and estimate of delay
25.2.2.2	Extension of time	Give particulars of effects and estimate of delay to NSCs
25.2.3	Extension of time	Give updating notices and copies to NSCs
25.3.4.1	Extension of time	Use best endeavours to prevent delay
25.3.4.2	Extension of time	Do all reasonably required to proceed with the Works
26.1.1	Loss and expense	Make application as soon as it is apparent regular progress is disturbed
26.1.2	Loss and expense	Submit information to support application under cl.26.1
26.1.3	Loss and expense	Submit details of loss and expense
26.4.1	Loss and expense	Pass on applications made by NSC
27.3.2	Determination by Employer	Immediately inform Employer if he becomes bankrupt, insolvent or similar

Clause No	Clause Title	Signpost
27.5.1	Determination by Employer	*Not* be bound to continue work from the date the Employer could first give notice
27.5.4	Determination by Employer	Allow and not hinder Employer's measures to secure site. Works and materials
27.6.2.1	Determination by Employer	Assign to the Employer benefit of supply or work agreements
27.6.3	Determination by Employer	Remove or have removed temporary buildings, plant, materials when required
28.4.1	Determination by Contractor	Remove temporary buildings, plant, materials and ensure Sub-Contractors do the same
28.4.3	Determination by Contractor	Prepare an account
28A.1.2	Determination by either	*Not* be entitled to give notice under cl.28A.1.1 where damage caused by own default
28A.3	Determination by either	Remove temporary buildings, plant, materials and ensure Sub-Contractors do the same
28A.5	Determination by either	Within 2 months of determination provide Employer with documents for an account
29.1	Works by Employer	Permit execution of work not forming part of the Contract
30.1.1.1	Certificates/payments	Include in an application for payment all relevant applications from NSCs
30.4A.2	Certificates/payments	On or before Date of Possession, provide and thereafter maintain a Retention Bond
30.4A.4	Certificates/payments	Arrange with the Surety for the aggregate sum to equate to Retention
30.6.1.1	Certificates/payments	Send all documents for adjustment of Contract Sum
30.8.5	Certificates/payments	Pay interest on late payment of monies due to Employer in Final Certificate
31.5.1.1	CIS	If Authorisation is a CIS 4, give Employer statement giving the direct cost of materials
31.5.2	CIS	Indemnify the Employer against any loss or expense arising from incorrect statements

CONTRACTOR SHALL …

Clause No	Clause Title	Signpost
31.7	CIS	Immediately inform Employer and present CIS 5 or 6 if any change from CIS 4 status
31.8	CIS	Immediately inform Employer if CIS 5 or 6 is withdrawn
31.11	CIS	Provide a CIS 24 voucher if Authorisation CIS 6 applies and payments were made
34.1.1	Antiquities	Use best endeavours not to disturb the object
34.1.2	Antiquities	Take steps to preserve the object
34.1.3	Antiquities	Inform the Architect or Clerk of Works of the discovery
Part 2		
35.2.1	General	Be permitted to tender for work carried out in the normal course of his business
35.2.1	General	*Not* sub-let such work without consent
35.5.1	Procedure	Make any reasonable objection at the earliest moment
35.7.1	Procedure	Complete and agree NSC/T Part 3
35.7.2	Procedure	Execute NSC/A
35.8	Procedure	Within 10 days of the instruction inform Architect of any failure to comply with cl.35.7
35.9.2	Procedure	Comply with cl.35.7 if Architect rejects grounds for non-compliance
35.13.2	Payment	Duly discharge each interim payment
35.13.3	Payment	Provide the Architect with reasonable proof of discharge

Clause No	Clause Title	Signpost
35.14.1	Extension of Period	*Not* extend NSC time unless through procedures of NSC/C
35.18.1.2	Early Final Payment	Pay the difference between substitute and recovered prices
35.19.1	Early Final Payment	Be responsible for loss up to Practical Completion of Works
35.21	Clause 2.1 of NSC/W	*Not* be responsible for NSC Works to which cl.2.1 of NSC/W relates
35.24.6.2	Re-nomination	Inform the Architect whether NSC employment determined
35.24.6.2	Re-nomination	Inform the Architect that NSC employment determined
35.25	Determination	*Not* determine NSC employment without instruction
36.5.2	Nominated Suppliers	*Not* be obliged to enter a contract until Architect has approved restrictions etc

Part 3

Clause No	Clause Title	Signpost
38.1.8	Fluctuations – taxes etc.	Be deemed to pay Employer's contributions for contracted-our employment
38.3.1	Sub-let/Domestics	Incorporate as Sub-Contract conditions the provisions of cl.38
38.4.1	Clause 38 provisions	Give written notice to the Architect of the occurrence of any of the events listed
38.4.2	Clause 38 provisions	Give notice under 38.4.1 within a reasonable time – a condition precedent to payment
38.4.5	Clause 38 provisions	Provide evidence and computations as soon as reasonably practicable
39.2.7	Fluctuations – costs, taxes etc	Be deemed to pay Employer's contributions for contracted-out employment
39.4.1	Sub-let/Domestics	Incorporate as Sub-Contract conditions the provisions of cl.39

CONTRACTOR SHALL ...

Clause No	Clause Title	Signpost
39.5.1	Clause 39 provisions	Give written notice to the Architect of the occurrence of any of the events listed
39.5.2	Clause 39 provisions	Give notice under 39.5.1 within a reasonable time – a condition precedent to payment
39.5.5	Clause 39 provisions	Provide evidence and computations as soon as reasonably practicable
40.3	Price adjustment formulae	Insert the market price of articles in a list attached to the Contract Bills
40.6.3	Price adjustment formulae	Operate relevant parts of cl.40 and Formula Rules during absence of publication
	Part 5	
42.2	Performance Specified Work	Provide a Contractor's Statement before carrying out any work
42.2	Performance Specified Work	Refer Contractor's Statement to Planning Supervisor before issue to the Architect
42.2	Performance Specified Work	Make Planning Supervisor's amendments before issue to Architect
42.13	Performance Specified Work	Provide an analysis of the work if not analysed in the Contract Bills
42.14	Performance Specified Work	Comply with instructions relative to integration of the PSW with the main Works
42.15	Performance Specified Work	Specify within 7 days any injurious affection arising from Architect's instruction
42.17	Performance Specified Work	Exercise reasonable skill and care in providing Performance Specified Work

Clause No	Clause Title	Signpost
	Appendices	
VAT 1A.2	VAT Agreement	Give written notice of the rate chargeable and of any change in the rate
VAT 1.1	VAT Agreement	Give a written provisional assessment of respective values
VAT 1.1.2	VAT Agreement	Specify rates of tax chargeable
VAT 1.2.2	VAT Agreement	Reply within three days to Employer's objection
VAT 1.3.1	VAT Agreement	Issue to the Employer a final written statement of value
VAT 1.3.1	VAT Agreement	Specify rates of tax chargeable
VAT 1.3.1	VAT Agreement	State total amount of tax already received
VAT 1.3.4	VAT Agreement	Refund excess payments within 28 days
VAT 1.3.4	VAT Agreement	Send a receipt at time of refund
VAT 1.4	VAT Agreement	Issue a receipt upon receipt of all amounts paid
VAT 2.2	VAT Agreement	Disregard deductions for liquidated damages
VAT 3.1	VAT Agreement	Make any requested application to the Commissioners
VAT 3.1	VAT Agreement	Make any necessary appeals against Commissioners' decisions
VAT 3.1	VAT Agreement	Account for any costs awarded in his favour
VAT 3.3	VAT Agreement	Refund within 28 days amounts adjudicated as overpaid

CONTRACTOR'S STATEMENT

Clause No	Clause Title	Signpost
	Part 1	
1.3	Definitions etc.	Contractor's Statement
2.4.1	Contractor's obligations	Discrepancies between Statement and any instruction
2.4.2	Contractor's obligations	Discrepancies within the Statement
6.1.6	Statutory obligations	Divergence between Statement and Statutory Requirements
8.1.1	Work, mats. and goods	Kinds and standards of goods and materials included in Statement
8.1.2	Work, mats. and goods	Standards of workmanship included in Statement
8.1.4	Work, mats. and goods	No substitution of materials or goods without consent
13.5.6.2	Variations/Prov. Sums	Valuation of additional or substituted work
	Part 5	
	Performance Specified Work	**Principal applicable clauses**

Clause No	Clause Title	Signpost
	Part 1	
1.3	Definitions etc.	Completion Date
1.3	Definitions etc.	Date for Completion
13A.2.2	Variation instruction	13A Quotation to include for time adjustments relative to the Date for Completion
13A.3.2.3	Variation instruction	Revised Completion Date may be earlier than Date for Completion
25.3.6	Extension of time	Decision under cls.25.3.2 or 25.3.3.2 shall not fix date earlier than Date for Completion
	Appendices	
1.3	Appendix	Date for Completion

DAYWORK

Clause No	Clause Title	Signpost
	Part 1	
13.5.4.1	Variations/Prov. Sums	Valuation of work which cannot properly be valued by measurement
13.5.4.2	Variations/Prov. Sums	Valuation of specialist work which cannot properly be valued by measurement
13.5.6.5	Variations/Prov. Sums	Valuation of additional or substituted work relating to PSW
	Part 3	
38.5.1	Clause 38 provisions	Cls.38.1 - .3 do not apply where daywork allowed
39.6.1	Clause 39 provisions	Cls.39.1 - .4 do not apply where daywork allowed

Clause No	Clause Title	Signpost
	Part 1	
30	Certificates/payments	**Principal applicable clause**
3	Contract Sum	To be taken into account in the next Interim Certificate as soon as amounts ascertained
4.1.2	Architect's instructions	Cost of employing others to carry out instructions
7	Levels and setting out	For errors not required to be amended
8.4.2	Work, materials and goods	If work, goods, materials, not in accordance with contract, are allowed to remain
13.4.1.2.A3	Contractor's Price Statement	Adjustment for following acceptance of Price Statement
13.7	Variations/Prov. Sums	Adjustment for Valuation under cl.13.4.1.1 or for a 13A Quotation
17.2; 17.3	Practical Compl./Defects	If Architect instructs otherwise than for Contractor to make good defects
18.1.4	Partial Possession	Proportion of liquidated damages
21.1.3	Insurance: persons etc	Cost of premiums where Contractor has failed to insure
22A.2	All Risks – by Contractor	Cost of premiums where Contractor has failed to insure
22FC.3	Joint Fire Code	Cost of employing others to carry out Remedial Measures
24.2.1	Damages	Liquidated damages
27.5.4	Determination by Employer	Cost of protecting materials if Contractor fails to do so
27.6.2.2	Determination by Employer	Direct payment by Employer of Contractor's suppliers or Sub-Contractors
28.4.2	Determination by Contractor	From Retention that may otherwise have become due for payment

DEDUCTIONS

Clause No	Clause Title	Signpost
28.4.3	Determination by Contractor	Amounts due to be paid to Contractor without deduction of Retention
28A.4	Determination by either	From Retention that may otherwise have become due for payment
28A.5.5	Determination by either	Amounts due to be paid to Contractor without deduction of Retention
	Part 2	
35.13.5.2	Payment	Amounts Contractor has failed to discharge
35.13.6.2	Payment	Credits for amounts paid prior to nomination
35.24.9	Re-nomination	Extra over NSC price resulting from further nomination
	Part 3	
38.4.4	Clause 38 provisions	Any amount which may become allowable – subject to cls.38.4.5 - .4.7
38.4.6	Clause 38 provisions	No deduction shall alter the amount of profit
38.4.7	Clause 38 provisions	No deduction if event otherwise causing the adjustment occurs after Completion Date
39.5.4	Clause 39 provisions	Any amount which may become allowable – subject to cls.39.5.5 - .5.7
39.5.6	Clause 39 provisions	No deduction shall alter the amount of profit
39.5.7	Clause 39 provisions	No deduction if event otherwise causing the adjustment occurs after Completion Date

Clause No	Clause Title	Signpost
	Appendices	
VAT 1.3.3	VAT Agreement	Balance of tax paid at final payment after deducting tax already paid
VAT 2.1; .2; .3	VAT Agreement	Deductions under cl.24 disregarded in calculating VAT
1	Retention Bond	Where Employer does not exercise right to deduct Retention
4(ii)	Retention Bond	State amount of Retention that would have been deductible
4(iii) (e)	Retention Bond	Identification the cost of items deductible under the Contract

DEEMED TO BE

Clause No	Clause Title	Signpost
	Part 1	
4.2	Architect's instructions	Instructions empowered by the Conditions specified
4.3.2.2	Architect's instructions	Confirmation of instructions complied with but unconfirmed
9.1	Royalties etc.	Included in the Contract Sum
10	Person-in-charge	Instructions given to the person-in-charge
12	Clerk of Works	Confirmation within two days of Clerk of Works direction
13.4.1.2.A4.2	Contractor's Price Statement	Non-acceptance of amended Price Statement
13.4.1.2.A5	Contractor's Price Statement	Non-acceptance of amended Price Statement
14.1	Contract Sum	Quantity and quality as set out in the Contract Bills
14.2	Contract Sum	Error in computation accepted by the parties
17.1	Practical Compl./Defects	Date of Practical Completion
17.4	Practical Compl./Defects	Date of Completion of Making Good Defects
18.1.1	Partial possession	Date of Practical Completion
18.1.1	Partial possession	Date of start of Defects Period
27.1	Determination by Employer	Receipt of notice when sent by special or recorded delivery
28.1	Determination by Contractor	Receipt of notice when sent by special or recorded delivery
28A.1.1	Determination by either	Receipt of notice when sent by special or recorded delivery

Clause No	Clause Title	Signpost
29.3	Works by Employer	Persons employed are the direct responsibility of the Employer
34.2	Antiquities	Employer responsible for a person dealing with the object

Part 2

35.2.1	Contractor tender	C1.13.3 items as being included in Contract Bill
35.2.1	Contractor tender	C1.13.3 items as being set out in the Appendix
35.16	Practical Completion	Date of Practical Completion
36.1.1.3	Nominated Suppliers	Expenditure of Provisional Sum in favour of sole supplier
36.1.1.4	Nominated Suppliers	Variation causing expenditure in favour of a sole supplier

Part 3

38.1	Fluctuations – taxes etc.	Contract Sum calculated in manner set out in the clause
38.1.8	Fluctuations – taxes etc.	Employer's contributions as if employment were not contracted out
38.2	Fluctuations – taxes etc.	Contract Sum calculated in manner set out in the clause
38.4.3	Clause 38 provisions	Quantity Surveyor and Contractor may agree net amount payable or allowable
39.1	Fluctuations – costs, taxes etc	Contract Sum calculated in manner set out in the clause
39.2	Fluctuations – costs, taxes etc	Contract Sum calculated in manner set out in the clause
39.2.7	Fluctuations – costs, taxes etc	Employer's contributions as if employment were not contracted out

DEEMED TO BE

Clause No	Clause Title	Signpost
39.3	Sub-let/Domestics	Contract Sum calculated in manner set out in the clause
39.5.3	Clause 39 provisions	Quantity Surveyor and Contractor may agree net amount payable or allowable
40.2	Price adjustment formulae	Amendment to clause 30
40.5	Price adjustment formulae	Quantity Surveyor and Contractor may agree alteration to methods and procedures

Part 4

41A.4.2	Adjudication	Receipt of notice when sent by special or recorded delivery

Appendices

4	Advance Payment Bond	Payment by the Surety as a valid payment
6	Off-site Materials etc. Bond	Payment by the Surety as a valid payment

Clause No	Clause Title	Signpost
	Part 1	
17	Practical Compl./Defects	**Principal applicable clause**
1.3	Definitions etc	Certificate of Making Good Defects
1.3	Definitions etc	Defects Liability Period
18.1.1	Partial Possession	Defects Liability Period commencement date
18.1.3	Partial Possession	Certificate to be issued when defects in relevant part are made good
27.6.1	Determination by Employer	Employer may employ others to carry out the works and make good defects
27.6.1	Determination by Employer	Employer may use existing plant, materials etc. and order additional
27.6.3.2	Determination by Employer	Cl.27.6.5 account to be prepared upon completion of making good defects
30.1.3	Certificates/payments	Issue of an Interim Certificate upon completion of making good defects
30.4.1.3	Certificates/payments	Timing of deduction of half Retention
30.8.1	Certificates/payments	Timing of issue of Final Certificate
	Part 2	
35.17.1	Final Payment	Remedying of defects before certification of Sub-Contract Sum
36.4.2	Nominated Suppliers	Nominated Supplier to make good defects appearing before end of DLP
36.4.2	Nominated Suppliers	Nominated Supplier to bear Contractor's expenses as a consequence of defects

DEFECTS

Clause No	Clause Title	Signpost
36.4.2.1	Nominated Suppliers	36.4.2 does not apply if reasonable examination would have revealed defects earlier
36.4.2.2	Nominated Suppliers	36.4.2 does not apply unless damage due to defective material or bad workmanship
	Appendices	
17.2	Appendix	Defects Liability Period
VAT 1.3.1	VAT Agreement	Timing of preparation of final statement of values
6(i)	Retention Bond	Issue of CCMGD is one of the occurrences releasing the Surety

Clause No	Clause Title	Signpost
	Part 1	
25	Extension of time	**Principal applicable clause**
5.4.1	Contract Documents	Any agreement to Information Release time Variation not to be unreasonably delayed
8.1.4	Work, materials and goods	Any consent to substitution of materials etc. not to be unreasonably delayed
16.1	Mats. and goods unfixed	Any consent to removal of materials etc. not to be unreasonably delayed
18.1	Partial Possession	Any consent of Contractor to partial possession not to be unreasonably delayed
19.2.2	Assignment/Sub-Contracts	Any consent to sub-letting not to be unreasonably delayed
19.3.2.1	Assignment/Sub-Contracts	Any consent for addition of name to list of named firms not to be unreasonably delayed
19.3.2.2	Assignment/Sub-Contracts	Any consent for addition of name to list of named firms not to be unreasonably delayed
19.4.2.1	Assignment/Sub-Contracts	Any consent to removal of sub-let firm's materials etc. not to be unreasonably delayed
22.D.1	Insurance: Loss of LADs	Any instruction for acceptance of insurance quotation not to be unreasonably delayed
23.3.2	Possession, completion	Any consent for Employer's early use of the site etc. not to be unreasonably delayed
27.5.4	Determination by Employer	Contractor not to delay Employer's measures to secure/protect materials and Works.
28.2.2.3	Determination by Contractor	By the Employer as a ground for determination
29.2	Works by Employer	Any consent for execution of Works by Employer not to be unreasonably delayed

DELAY

Clause No	Clause Title	Signpost
	Part 2	
35.1.4	General	Any agreement to choice of a nominated firm not to be unreasonably delayed
35.18.1.2	Early Final Payment	Any agreement to substituted Sub-Contractor price not to be unreasonably delayed
	Part 3	
40.6.1	Price adjustment formulae	Valuation if publication of Monthly Bulletins delayed
40.6.2	Price adjustment formulae	Valuation if publication of Monthly Bulletins resumed before issue of Final Certificate
40.6.3	Price adjustment formulae	Valuation where publication of Monthly Bulletins delayed
	Part 5	
42.15	Performance Specified Work	Any consent to an instruction affecting PSW not to be unreasonably delayed
42.16	Performance Specified Work	No extension of time if cause of delay is late receipt of Contractor's Statement
	Appendices	
5	Retention Bond	Any consent by Surety to assignment not to be unreasonably delayed

Clause No	Clause Title	Signpost
Art. 7A	Articles	Dispute arising after determination
Art. 7B	Articles	Dispute arising after determination
	Part 1	
27; 28; 28A	Determination	**Principal applicable clauses**
19.4.1	Assignment/Sub Contracts	Determination of employment of a Domestic Sub-Contractor
19.4.3	Assignment/Sub Contracts	Late payment interest does not affect Sub-Contractor's rights of determination
20.3.1	Insurance: persons etc.	Cl.20.2 references to property do not include Works etc before date of determination
22.3.1	Insurance of the Works	Recognition of NSC within Joint Names Policy ceases at date of determination
22.3.2	Insurance of the Works	Joint Names Policy recognition of Domestic Sub-Contractors stops at determination
22A.1	All Risks – by Contractor	Insurance ceases at date of determination
22A.5.2.2	All Risks – by Contractor	Determination in the event of an act of terrorism
22B.1	All Risks – by Employer	Insurance ceases at date of determination
22C.1	Insurance - existing structures	Insurance ceases at date of determination
22C.1A.3	Insurance - existing structures	Determination in the event of an act of terrorism
22C.2	Insurance - existing structures	Insurance (existing structures) ceases at date of determination

DETERMINATION

Clause No	Clause Title	Signpost
22C.2	Insurance - existing structures	Insurance (Works) ceases at date of determination
22C.4.3.1	Insurance - existing structures	Determination may follow loss or damage, if fair and reasonable
22C4.3.2	Insurance - existing structures	Procedure for determination under these clauses
22C.4.4	Insurance - existing structures	Procedure if no notices of determination are issued
22C.5.2	Insurance - existing structures	Employer may give notice of determination following acts of terrorism
30.1.1.2	Certificates/payments	Interest for late payment does not affect Contractor's rights of determination

Part 2

Clause No	Clause Title	Signpost
35.24; 35.25	Re-nomination; Determination	**Principal applicable clauses**
36.4.5	Nominated Suppliers	No obligation to deliver after determination

Appendices

Clause No	Clause Title	Signpost
VAT 8	VAT Agreement	VAT payable by reason of determination
1.4.1	EDI	Any determination must be in writing
4(iii) (d)	Retention Bond	Costs of determination as a ground upon which demand may be made on Surety

Clause No	Clause Title	Signpost
	Part 1	
26	Loss and expense	**Principal applicable clause**
8.4.3	Work, materials and goods	No addition to Contract Sum for non-compliant work made the subject of a Variation
8.4.4	Work, materials and goods	No addition to Contract Sum for cost of checking for further non-compliant work
8.5	Work, materials and goods	No addition to Contract Sum for cost of instructions requiring compliance with cl.8.1.3
13.4.1.2.A1.1	Contractor's Price Statement	Price Statement to attach assessment of related requirements under cl.26
13.4.1.2.A7.1.1	Contractor's Price Statement	Within 21 days of Price Statement receipt, QS to notify Contractor of cl.26 decision
13.4.1.2.A7.2	Contractor's Price Statement	Cl.26 to apply in the absence of notification from QS of cl.26 decision
13.5	Variations/Prov. Sums	No allowance where reimbursable under other provisions
13A.2.3	Variation instruction	13A Quotation to separately identify cl.26 ascertainment
13A.8	Variation instruction	QS to assess direct loss/expense in evaluating any Variation to a 13A Quotation item

DIRECT LOSS/EXPENSE

Clause No	Clause Title	Signpost
27.6.3	Determination by Employer	Employer not responsible when he removes and sells property
27.6.5.1	Determination by Employer	Direct loss recoverable by the Employer in the event of determination
27.7.1.2	Determination by Employer	Statement to include amount of direct loss where Employer decides not to complete
28.4.3.2; .4	Determination by Contractor	Contractor's account to include amount of any recoverable direct loss
28A.5.2; .5	Determination by either	Employer's account to include amount of any recoverable direct loss
30.2.2.2	Certificates/payments	Amounts under cl.26.1 are not subject to Retention
30.6.1.2.1	Certificates/payments	Timing for ascertaining amounts under cls.26.1 + 26.4.1
30.6.2.13; .17	Certificates/payments	Amounts ascertained under cl.26.1 added to Contract Sum
30.9.1.4	Certificates/payments	Final Certificate as conclusive evidence of settlement of cl.26 entitlements
34.3.1	Antiquities	To be ascertained by the Architect or QS
34.3.2	Antiquities	Statement of extension of time if necessary for ascertainment
34.3.3	Antiquities	Amounts ascertained under cl.34.3 added to Contract Sum

Clause No	Clause Title	Signpost
	Part 2	
35.26.1	Determination	Architect to provide Contractor with amount of direct loss incurred by the Employer
	Part 5	
42.16	Performance Specified Work	Clause 26 of no effect where delay is caused by late receipt of Contractor's Statement
	Appendices	
26.1	Appendix	Deferment of Date for Possession

DISCOUNT

Clause No	Clause Title	Signpost
	Part 1	
30.6.2.8	Certificates/payments	In adjustment of Contract Sum amounts for Nominated Suppliers to include discount
	Part 2	
36.3.1	Nominated Suppliers	Amounts chargeable to Employer are net of discount other than cl.36.4.4 discount
36.3.1.3	Nominated Suppliers	Price adjustments are net of discount other than cl.36.4.4 discount
36.4.4	Nominated Suppliers	Contractor is allowed a 5% cash discount for prompt payment
36.4.5	Nominated Suppliers	Delivery, after determination, of goods already paid
36.4.6	Nominated Suppliers	Contractor is allowed a 5% cash discount for prompt payment

Clause No	Clause Title	Signpost
Art. 3	Articles	Objection to a replacement Architect
Art. 4	Articles	Objection to a replacement Quantity Surveyor
Art. 5	Articles	Reference to adjudication
Art. 7A	Articles	Enforcement of a decision by the Adjudicator not subject to Arbitration
Art. 7B	Articles	Reference of disputes to legal proceedings

Part 1

4.2	Architect's Instructions	Power of Architect to issue instructions
13.4.1.2.A4.3	Contractor's Price Statement	Non-acceptance of amended Price Statement
13.4.1.2.A5	Contractor's Price Statement	Non-acceptance of Price Statement
20.3.1	Injury and indemnity	Reference to 'property real or personal' in the context of validity of determination
22C.4.3.1	Insurance - existing structures	Validity of determination following loss or damage
30.9.4	Certificates/payments	Timing for commencement of arbitration or legal proceedings after Final Certificate
31.14	CIS	Dispute settlement procedures are as in the Contract unless otherwise stated

Part 2

36.4.8	Nominated Suppliers	Clause 41B applies in any dispute with the Contractor referred to arbitration

DISPUTES/DIFFERENCES

Clause No	Clause Title	Signpost
	Part 4	
41A	Adjudication	
41B	Arbitration	**Principal applicable clauses**
41C	Legal proceedings	
	Appendices	
Art. 7A; 7B	Appendix	Settlement of disputes
4	Advance Payment Bond	Payments due notwithstanding any dispute
6	Off-site Materials etc. Bond	Payments due notwithstanding any dispute
VAT 5	VAT Agreement	VAT Agreement applies to any payment adjustment arising from an award
1.4.4	EDI	Invoking of dispute resolution measures must be in writing and not on electronic data
2	EDI	Dispute settlement procedures on an EDI matter are as in the Contract

Clause No	Clause Title	Signpost
Art. 1	Articles	Contractor to carry out Works shown on Contract Documents
	Part 1	
5	Contract Documents	**Principal applicable clause**
1.3	Definitions etc	Contract Documents
1.3	Definitions etc	Documents annexed to NSC/A and NSC/C
1.3	Definitions etc	Numbered Documents
1.3	Definitions etc	Works as referred to in the Contract Documents
2.1	Contractor's obligations	Drawings, Bills, Articles, Conditions
2.3	Contractor's obligations	Discrepancies between documents to be notified to the Architect
6.1.2	Statutory obligations	Note of divergence between documents/Statutory Requirement
6.1.3	Statutory obligations	Architect to issue instructions on divergence between documents
6.1.4.3	Statutory obligations	Position where divergent work becomes the subject of an emergency
6.1.5	Statutory obligations	No liability if work carried out to Contract Documents etc.
13.4.1.2.A1	Contractor's Price Statement	Price Statement for work where an Approximate Quantity is in Contract Documents
13.4.1.3	Variations/Prov. Sums	Valuation of NSC works under an Approximate Quantity in Numbered Documents

DOCUMENTS

Clause No	Clause Title	Signpost
13.5.6.1	Variations/Prov. Sums	Valuation of Performance Specified Work to include work in preparation of documents
26.2.3	Loss and Expense	Discrepancies between drawings, bills and Numbered Documents as a matter
28.2.2.4	Determination by Contractor	Suspension of Works because access according to Contract Documents unavailable
28A.5	Determination by either	Contractor to provide documents for preparation of account
30.6.1.1	Certificates/payments	Documents for adjustment of the Contract Sum including NSC documents
30.6.1.2	Certificates/payments	Final amounts to be ascertained within 3 months of receipt of documents
30.9.1.1	Certificates/payments	Final Certificate conclusive evidence of quality accords with Numbered Documents
	Part 2	
35.2.2	General	Clause 13 shall apply to NSC work as if there were equivalent references in documents
35.4	Procedure	Document references
35.6.1	Procedure	Documents to accompany nomination instruction
35.17.2	Early Final Payment	Necessary for final adjustment of Sub-Contract Sum must have been sent by NSC
35.21.3	Clause 2.1 of NSC/W	Contractor not responsible for Performance NSC work in numbered tender documents
	Part 5	
42.2	Performance Specified Work	Contractor to provide document 'Contractor's Statement' before carrying out PSW

Clause No	Clause Title	Signpost
	Appendices	
2.8	Clause 8.4.4 Code of Practice	Contractor's failure to carry out tests specified in Contract Documents
1.2	EDI	Types of communication and persons to which EDI applies as Contract Documents
1.3	EDI	Message Standards are as in Contract Documents
1.4	EDI	EDI suffices as fulfilling requirements for items to be in writing with some exceptions
7	Retention Bond	Demand to Surety must be accompanied by documents specified in cl.4

DOMESTIC SUB-CONTRACTOR

Clause No	Clause Title	Signpost
	Part 1	
19.2 - 19.4	Assignment/Sub Contracts	**Principal applicable clauses**
1.3	Definitions etc.	Domestic Sub-Contractor
6.3	Statutory obligations	Status of local authorities or statutory undertakers working on the site
6A.4	CDM Regulations	Sub-Contractors are to provide information to the Planning Supervisor
11	Access to the Works	Access for Architect to workshops of Domestic SC
22.3.2	Insurance of the Works	Joint Names Policies are to recognise Domestic Sub-Contractors
22A.4.4	All risks - Contractor	Insurance money for Domestic Sub-Contractors authorised to be paid to the Employer
22A.5.3	All risks - Contractor	Amounts payable are not reduced by reason of contributory acts
22B.3.4	All risks - Employer	Insurance money for Domestic Sub-Contractors authorised to be paid to the Employer
22B.4.3	All risks - Employer	Amounts payable are not reduced by reason of contributory acts
22C.4.2	Insurance - existing structures	Insurance money for Domestic Sub-Contractors authorised to be paid to the Employer
22C.5.3	Insurance - existing structures	Amounts payable are not reduced by reason of contributory acts
27.6.2.2	Determination by Employer	Direct payment by Employer
28.4	Determination by Contractor	Consequences of determination under cl.28.2 or .3
28.4.1	Determination by Contractor	Removal of the property of Sub-Contractors
28A.3	Determination by either	Removal of the property of Sub-Contractors

Clause No	Clause Title	Signpost
29.3	Works by Employer	Persons employed under cls.29.1 and .2 are not Sub-Contractors
	Part 3	
38.1.4	Fluctuations – taxes etc.	References to craftsman rates to include employees of Domestic Sub-Contractors
38.3.1	Sub-let/Domestics	Domestic Sub-Contracts to include provisions with like effect as cl.38
38.4.5	Clause 38 provisions	Evidence for computations to be submitted on behalf of Domestic Sub-Contractors
39.1.4	Fluctuations – costs, taxes etc	References to craftsman rates to include employees of Domestic Sub-Contractors
39.4.1	Sub-let/Domestics	Domestic Sub-Contracts to include provisions with like effect as cl.39
39.5.5	Clause 39 provisions	Evidence for computations to be submitted on behalf of Domestic Sub-Contractors
40.5	Price adjustment formulae	Adjustments under cl.40.5 have no effect on adjustments payable to a Sub-Contractor
	Appendices	
2.6	Code of Practice: Clause 8.4.4	Relevant records of a Sub-Contractor to be taken into account

DRAWINGS

Clause No	Clause Title	Signpost
Recitals	Articles	The Employer has caused drawings to be prepared
Recitals	Articles	Drawings numbered in the Articles are 'Contract Drawings'
	Part 1	
5	Contract Documents	**Principal applicable clause**
1.3	Definitions etc.	Contract Drawings
2.3.1	Contractor's obligations	Discrepancies to be notified to the Architect
2.3.4	Contractor's obligations	Discrepancies to be notified to the Architect
7	Levels and setting out	Drawings to be supplied for accurate setting out
13.5.6.1	Variations/Prov. Sums	Valuation of Performance Specified Work to include work in preparation of drawings
17.1	Practical Compl./Defects	Supply of as-built drawings under cl.5.9 before issue of CPC
25.4.12	Extension of time	Failure to give access as a Relevant Event
26.2.3	Loss and expense	Discrepancies with Contract Bills as a matter under cl.26.1
26.2.6	Loss and expense	Failure to give access as a matter under cl.26.1
30.9.1.1	Certificates/payments	Final Certificate conclusive evidence of quality accords with Drawings

Clause No	Clause Title	Signpost
	Part 2	
35.1.3.1	General	Variations for work additional to that in Contract Bills
35.2.2	General	References within tender accepted by Contractor
	Part 5	
42.1.3	Performance Specified Work	Part of definition as shown on Contract Drawings

EMPLOYER MAY...

Clause No	Clause Title	Signpost
Art. 5	Articles	Require a dispute to be referred to a different Arbitrator
	Part 1	
1.9	Definitions etc.	Give notice of appointment of Employer's Representative
4.1.2	Architect's instructions	Employ others to execute work Contractor has failed to do
18.1	Partial possession	Take possession of part of the Works
18.1.4	Partial possession	Give notice of deductions to be made under cl.30.1.1.4
19.1.2	Assignment/Sub-Contracts	Assign the right to bring proceedings
21.1.2	Insurance: persons etc	Require production of policies or premium receipts
21.1.3	Insurance: persons etc	Insure for cl.21.1 risks when Contractor fails to do so and deduct cost from monies due
21.2.4	Insurance: persons etc	Insure for cl.21.2 risks when Contractor fails to do so
22A.2	All risks - Contractor	Insure in joint names when Contractor fails to do so and deduct cost from monies due
22A.3.1	All risks - Contractor	Require production of independent policies or premium receipts
22C.4.3.1	Insurance - existing structures	Determine employment of Contractor following loss or damage
22C.4.3.1	Insurance - existing structures	Invoke relevant procedures to decide whether determination was just and equitable
22D.4	Insurance: loss of LADs	Insure for cl.22D.1 insurance when Contractor fails to do so

Clause No	Clause Title	Signpost
22FC.3.2	Joint Fire Code	Employ others to carry out Remedial Measures if Contractor fails to start or proceed
23.1.2	Possession, completion	Defer giving possession
23.3.2	Possession, completion	Use or occupy the site or a part with consent of the Contractor
24.2.1.1	Damages	Require a sum as liquidated damages
24.2.1.1	Damages	Recover liquidated damages as a debt
24.2.1.2	Damages	Deduct liquidated damages from monies due
27.2.2	Determination by Employer	Determine employment if Contractor continues default
27.2.3	Determination by Employer	Determine employment if Contractor repeats default
27.3.3	Determination by Employer	Agree to reinstatement of Contractor's employment
27.3.4	Determination by Employer	Determine employment if Contractor becomes insolvent etc.
27.4	Determination by Employer	Determine employment if Contractor acts corruptly
27.5.3	Determination by Employer	Make interim arrangements for work to be carried out
27.5.4	Determination by Employer	Take reasonable measures to protect the site, works and materials and charge costs
27.6.1	Determination by Employer	Employ and pay others to complete the Works
27.6.2.2	Determination by Employer	Pay suppliers or Sub-Contractors direct and charge to the Contractor
27.6.3	Determination by Employer	Remove and sell property of the Contractor holding proceeds to his credit
29.2	Works by Employer	Arrange for the execution of Works not part of the Contract
30.1.1.2	Certificates/payments	Exercise right to deduct from monies due

EMPLOYER MAY ...

Clause No	Clause Title	Signpost
30..4.1	Certificates/payments	Deduct Retention
30.4A.4	Certificates/payments	Deduct Retention where the otherwise due amount exceeds provision of Bond
30.8.3	Certificates/payments	Inform Contractor in writing of any proposed deductions
30.9.2.1; .2	Certificates/payments	Status of Final Certificate if either party commences proceedings before issue
30.9.3	Certificates/payments	Status of Final Certificate if either party commences proceedings after issue
30.9.4	Certificates/payments	Commence arbitration or legal proceedings within 28 days of Adjudicator's decision
31.12	CIS	Correct errors in calculations

Part 2

Clause No	Clause Title	Signpost
35.13.6.2	Payment	Make adjustments to take account of monies paid to NSC before nomination
35.24.9	Re-nomination	Deduct extra cost of further nomination from monies due
35.24.9	Re-nomination	Recover extra cost of further nomination as a debt

Clause No	Clause Title	Signpost
	Part 4	
41A	Adjudication	Several references to matters at the discretion of both parties
41B	Arbitration	Several references to matters at the discretion of both parties
	Appendices	
	Appendix	Appointment of Adjudicator
5	Advance Payment Bond	Actions not needing consent of Surety
7	Off-site Materials Bond	Actions not needing consent of Surety
VAT 1A.4	VAT Agreement	Give notice stating that cl.1A does not apply
VAT 3.1	VAT Agreement	Request the Contractor to obtain a Commissioners' decision

EMPLOYER SHALL ...

Clause No	Clause Title	Signpost
Art. 3	Articles	Nominate a replacement Architect
Art. 4	Articles	Nominate a replacement Quantity Surveyor
Art. 6.1	Articles	Nominate a replacement Planning Supervisor
Art. 6.2	Articles	Nominate a replacement Principal Contractor
	Part 1	
1.6	Definitions etc.	Notify Contractor if he replaces Planning Supervisor or Principal Contractor
5.7	Contract Documents	*Not* divulge rates or prices in the Contract Bills
6A.1	CDM Regulations	Ensure the Planning Supervisor and Principal Contractor carry out duties
6A.2	CDM Regulations	Notify Planning Supervisor and Architect of amendments to H&S Plan
13A.3.1	Variation instruction	Notify Contractor in Writing if he wishes to accept a 13A Quotation
19.1 .1	Assignment/Sub-Contracts	*Not* assign the Contract without written consent
19.3.2.1	Assignment/Sub-Contracts	Be entitled, with consent, to add names to lists of named specialist firms
19.3.2.2	Assignment/Sub-Contracts	Add, by agreement, names to lists of specialist firms if less than three persons named
22.2.1	Insurance of the Works	Where cl.22A applies, ensure Joint Names Policy includes items specific to NSCs
22A.4.4	All risks - Contractor	Pay insurance monies to the Contractor by instalments
22A.5.1	All risks - Contractor	Notify Contractor immediately if insurers withdraw terrorism cover

Clause No	Clause Title	Signpost
22A.5.2	All risks - Contractor	Notify the Contractor of chosen procedure in the event of damage by terrorism
22A.5.3	All risks - Contractor	*Not* reduce payments to the extent the Contractor may have contributed to damage
22B.1	All risks - Employer	Take out and maintain a Joint Names all risks policy
22B.1	All risks - Employer	Maintain Joint Names policy until issue of CPC or date of determination
22B.2	All risks - Employer	Provide documentary evidence of insurance
22B.4.1	All risks - Employer	Notify Contractor immediately if insurers withdraw terrorism cover
22B.4.2	All risks - Employer	Notify the Contractor of chosen procedure in the event of damage by terrorism
22B.4.3	All risks - Employer	*Not* reduce payments to the extent the Contractor may have contributed to damage
22C.1	Insurance - existing structures	Take out and maintain a Joint Names all risks policy – existing buildings
22C.1.A.1	Insurance - existing structures	Notify Contractor immediately if insurers withdraw 22C.1 terrorism cover
22C.1.A.1	Insurance - existing structures	Notify the Contractor of chosen procedure in the event of damage by terrorism
22C.1.A.2	Insurance - existing structures	Continue to require the Works to be carried out where clause 22C.1A.1.1 applies
22C.2	Insurance - existing structures	Take out and maintain a Joint Names all risks policy – the Works
22C.2	Insurance - existing structures	Maintain Joint Names policy until issue of CPC or date of determination
22C.3	Insurance - existing structures	Provide documentary evidence of insurance
22C.5.1	Insurance - existing structures	Notify Contractor immediately if insurers withdraw terrorism cover
22C.5.2	Insurance - existing structures	Notify the Contractor of chosen procedure in the event of damage by terrorism

EMPLOYER SHALL ...

Clause No	Clause Title	Signpost
22C.5.3	Insurance - existing structures	*Not* reduce payments to the extent the Contractor may have contributed to damage
22FC.2.1	Joint Fire Code	Comply with Joint Fire Code
22FC.4	Joint Fire Code	Indemnify the contractor against consequences of breaches of Joint Fire Code
23.3.1	Possession, completion	*Not* be entitled to take possession of any part before issue of CPC subject to cl.18
23.3.2	Possession, completion	Notify insurers if he occupies any part of Works prior to issue of CPC
24.2.2	Damages	Repay liquidated damages if Completion Date fixed later
27.4	Determination by Employer	Be entitled to determine Contractor's employment in event of corruption
27.7.1	Determination by Employer	Notify Contractor within 6 months of determination of a decision to abandon Works
27.7.1	Determination by Employer	Send Contractor a statement of account following 27.7.1 decision
28.3.2	Determination by Contractor	Notify Contractor if he becomes insolvent or similar
28.4.2	Determination by Contractor	Pay Retention deducted
28.4.3	Determination by Contractor	Pay the Contractor amounts properly due
28A.4	Determination by either	Pay half of Retention deducted
28A.5	Determination by either	Prepare an account
28A.5.5	Determination by either	Pay the Contractor amounts properly due
28A.7	Determination by either	Inform Contractor of amounts included within account for NSCs
28A.7	Determination by either	Notify NSCs of amounts included within account

Clause No	Clause Title	Signpost
30.1.1.1	Certificates/payments	Pay interest to the Contractor on delayed payments
30.1.1.3	Certificates/payments	Notify within 5 days amount he proposes to pay against an Interim Certificate
30.1.1.5	Certificates/payments	Pay the amount stated in the Certificate if he does not otherwise notify
30.3	Certificates/payments	Annex to the Contract Bills a list of items eligible for pre-delivery payment
30.4A.3	Certificates/payments	Release Retention held before production of Retention Bond
30.4A.5	Certificates/payments	Call on Retention Bond before Performance Bond if latter exists
30.5.3	Certificates/payments	Place Retention in a separate banking account if required
30.5.3	Certificates/payments	Certify to Architect if Retention placed in a separate account
30.5.3	Certificates/payments	Be entitled to benefit of any interest that accrues in separate account
30.5.4	Certifcates/payments	Inform the Contractor of amounts of deduction
30.8.2	Certifcates/payments	Notify within 5 days amount he proposes to pay against Final Certificate
30.8.4	Certifcates/payments	Pay the amount stated in the Certificate if he does not otherwise notify
30.8.5	Certifcates/payments	Pay interest to the Contractor on delayed payments
31.2	CIS	Inform the Contractor if he becomes a 'contractor'
31.3	CIS	*Not* make payment unless a valid Authorisation is provided
31.4.1	CIS	Notify the Contractor if not satisfied with validity of the Authorisation
31.4.2	CIS	*Not* make payment until he is satisfied with the Authorisation

EMPLOYER SHALL ...

Clause No	Clause Title	Signpost
31.5.1.2	CIS	Make statutory deduction from that part of the payment not the direct cost of materials
31.5.3	CIS	Make a fair estimate of the direct cost of materials
31.6	CIS	Pay any amount due without making the statutory deduction where CIS5 or 6 valid
31.8	CIS	Make no further payments if CIS5 or 6 withdrawn
31.9	CIS	Make no further payments until new Authorisations in place, if CIS5 or 6 expires
31.10	CIS	Provide a CIS25 where CIS4 applies
31.11	CIS	Add his tax reference to CIS24 and send the voucher to the Inland Revenue
31.11	CIS	Send copy of last item to Contractor
	Part 2	
35.13.5.2	Payment	Pay NSC amounts which Contractor has failed to discharge
35.13.5.2	Payment	*Not* be obliged to pay amounts in excess of reduction
35.13.5.3.3	Payment	Pay NSCs pro-rata to amounts available where necessary
35.13.6.1	Payment	Send Contractor statement of pre-nomination payments to be credited
35.18.1.2	Early Final Payment	Take steps to recover price of substituted Sub-Contractor

Clause No	Clause Title	Signpost
	Part 3	
40.6.3	Price adjustment formulae	Operate relevant parts of cl.40 and Formula Rules during absence of publication
	Appendices	
3(b)	Advance Payment Bond	Provide Surety with completed notice in form of attached Schedule
4	Off-site materials Bond	Provide Surety with completed notice in form of attached Schedule
VAT 1	VAT Agreement	Pay to the Contractor any VAT properly chargeable
VAT 1.2.1	VAT Agreement	Calculate and pay VAT upon receipt of assessment
VAT 1.2.2	VAT Agreement	Notify grounds for reasonable objection within 3 days
VAT 1.3.3	VAT Agreement	Calculate and pay balance upon receipt of final statement
VAT 1.3.4	VAT Agreement	Notify Contractor if VAT specified exceeds amounts paid
VAT 2.3	VAT Agreement	Pay tax to which cl.1A refers if that clause operates
VAT 3.2	VAT Agreement	Pay the full amount of VAT where necessary before appeal
VAT 3.3	VAT Agreement	Pay within 28 days any amounts adjudicated underpaid
VAT 4	VAT Agreement	Be discharged from further liability after payments of amounts due on Final Certificate
VAT 7	VAT Agreement	*Not* be obliged, in certain conditions of absence of receipt, to make further payments
2(i)	Retention Bond	Be entitled to receive sum demanded from Surety
5	Retention Bond	Be entitled to assign Bond with consent of Surety

ERROR

Clause No	Clause Title	Signpost
	Part 1	
2.2.2.2	Contractor's obligations	Error in the Contract Bills shall not vitiate the Contract
7	Levels and setting out	Contractor shall amend his own errors at his own cost
7	Levels and setting out	Architect may instruct errors not to be rectified and an appropriate deduction made
14.2	Contract Sum	Error in computation of Contract Sum is deemed accepted
21.2.1.2	Insurance: persons etc.	Insurance does not apply to damage due to design error
30.9.1.2	Certificates/payments	Effect of error on the status of Final Certificate
31.12	CIS	Correction of error in calculation of statutory deduction
	Part 5	
42.10	Performance Specified Work	Correction of error in information given Contract Bills

EXTENSION OF TIME

Clause No	Clause Title	Signpost
26.3	Loss and expense	Extensions under cl.25 to be stated if necessary to ascertain loss and expense
30.9.1.3	Certificates/payments	Final Certificate as evidence that extensions of time due under cl.25 are given
34.3.2	Antiquities	Extensions under cl.25 to be stated if necessary to ascertain loss and expense
	Part 2	
35.14	Extension of Period	**Principal applicable clause**
35.15.1	Failure to complete	Failure to complete within extended time
	Part 3	
38.4.8.1	Clause 38 provisions	Cl.38.4.7 applies only if cl.25 unamended, forms part of the Conditions
38.4.8.2	Clause 38 provisions	Cl.38.4.7 applies only if procedures for fixing new Completion Dates were observed
39.5.8.1	Clause 39 provisions	Cl.39.5.7 applies only if cl.25 unamended, forms part of the Conditions
39.5.8.2	Clause 39 provisions	Cl.39.5.7 applies only if procedures for fixing new Completion Dates were observed
40.7.2.1	Price adjustment formulae	Cl.40.7.1 applies only if cl.25 unamended, forms part of the Conditions
40.7.2.2	Price adjustment formulae	Cl.40.7.1 applies only if procedures for fixing new Completion Dates were observed

Clause No	Clause Title	Signpost
	Part 4	
41A.5.3	Adjudication	Extension of period within which Adjudicator must reach his decision
41B.1.1	Arbitration	Extension of period within which parties may agree the name of Arbitrator
	Part 5	
42.16	Performance Specified Work	Extension may not be given if cause of delay is late receipt of Contractor's Statement
42.16	Performance Specified Work	Extension may be given in respect of cl.25.4.15
	Appendices	
5(c)	Advance Payment bond	Extensions of time may be granted without consent of Surety
7(c)	Off-site Materials etc. Bond	Extensions of time may be granted without consent of Surety

FLUCTUATIONS

Clause No	Clause Title	Signpost
	Part 1	
20	Injury and indemnity	**Principal applicable clause**
6.2	Statutory obligations	By Contractor for fees demandable by Act of Parliament etc.
9.1	Royalties etc.	By Contractor for infringement of patent rights
18.1	Partial possession	Identification of part of the Works taken into possession
21.1.1.1	Insurance: persons etc.	Taking out insurance is without prejudice to liability to indemnify under cl.20.
21.1.1.2	Insurance: persons etc.	Extent of indemnity for all claims other than for personal injury or death
21.2.1	Insurance: persons etc.	Appendix specifies amount of indemnity
21.2.1.1	Insurance: persons etc.	Insurance under cl.21.2.1 excludes Contractor liabilities under cl.20.2
21.3	Insurance: persons etc.	No obligation on Contractor to indemnify against Excepted Risks
22FC.4	Joint Fire Code	Each party to indemnify the other in respect of breaches of Joint Fire Code
28.4	Determination by Contractor	Cls.28.4.1, .2, .3 apply without prejudice to liabilities mentioned in cl.20.
28.4.1	Determination by Contractor	Prevention of injury/damage during removal of buildings etc.
28A.3	Determination by either	Prevention of injury/damage during removal of buildings etc.
31.5.2	CIS	Loss and expense caused by any incorrect statement
34.2	Antiquities	Persons working on antiquities are Employer's responsibility for purposes of cl.20

INDEMNITY

Clause No	Clause Title	Signpost
	Appendices	
VAT 3.1	VAT Agreement	Costs and expenses arising from an appeal

Clause No	Clause Title	Signpost
Art. 3	Articles	Architect may not overrule instructions etc. of a predecessor
	Part 1	
4	Architect's instructions	**Principal applicable clause**
2.2.2.2	Contractor's obligations	Correcting error etc. in Contract Bills treated as Variation
2.3	Contractor's obligations	Divergence between documents listed requires an instruction
2.3.3	Contractor's obligations	Divergence between instruction requiring a Variation and other Documents
2.4.1	Contractor's obligations	Divergence between Statement for PSW and an instruction
5.4.2	Contract documents	Issue of instructions as and when further information necessary
5.9	Contract documents	Expenditure of Provisional Sum for Performance Specified Work.
6.1.2	Statutory obligations	Divergence between cl.13.2 and Statutory Requirements
6.1.3	Statutory obligations	Timing of notification of divergence
6.1.3	Statutory obligations	C1.6.1.3 instructions requiring varied work to be treated as being under cl.13.2
6.1.4.1	Statutory obligations	Activity if emergency work is done before receipt of cl.6.1.1. instructions
6.1.4.3	Statutory obligations	Emergency work arising out of divergence

INSTRUCTIONS

Clause No	Clause Title	Signpost
6.1.5	Statutory obligations	Position of liability if work carried out to instruction etc.
6.1.6	Statutory obligations	Amending divergence between Statutory Requirements and Contractor's Statement
6.1.6	Statutory obligations	Compliance under this clause is subject to cl.42.15
6.1.7	Statutory obligations	Change in Statutory Requirement after Base Date as it affects PSW
7	Levels and setting out	Responsibility for errors
8.3	Materials, goods etc	Opening up work for inspection
8.4.1	Materials, goods etc	Removal from site of any non-compliant work, materials or goods
8.4.3	Materials, goods etc	Issued as a Variation as a consequence of instructions under cl.8.4.1
8.4.4	Materials, goods etc	Opening up works to check for further non-compliance
8.5	Materials, goods etc	As a consequence of any failure to comply with cl.8.1.3
8.5	Materials, goods etc	No addition to Contract Sum for instructions issued under this clause.
8.6	Materials, goods etc	Exclusion of any person employed
9.2	Royalties etc	Payment where arising out of instructions
10	Person-in-charge	Instructions to person-in-charge

Clause No	Clause Title	Signpost
12	Clerk of Works	Effect of directions by Clerk of Works
12	Clerk of Works	Confirmation of directions by Clerk of Works
13.2.1	Variations/Prov. Sums	Requiring a Variation
13.2.2	Variations/Prov. Sums	Instructions under 13.2.1 subject to Contractor's cl.4.1.1 right of reasonable objection
13.2.3	Variations/Prov. Sums	Valuation of Variations instructed under cl.13.2.1
13.2.3	Variations/Prov. Sums	Valuation of Variations instructed under cl.13A
13.2.3	Variations/Prov. Sums	Procedure if Contractor disagrees that cl.13A should apply
13.2.4	Variations/Prov. Sums	Sanctioning a Variation
13.3	Variations/Prov. Sums	Expenditure of Provisional Sums
13.4.1	Variations/Prov. Sums	Valuation by the QS
13.4.1.1	Variations/Prov. Sums	Valuation of Variations, Provisional Sums or work with Approximate Quantities.
13.4.1.2.A1	Contractor's Price Statement	Preparation of Price Statement for evaluation of work instructed under cl.13.4.1.1
13.4.1.3	Variations/Prov. Sums	Valuation of work by NSCs
13.4.2	Variations/Prov. Sums	Valuation of work the subject of a Provisional Sum
13.5.3.3	Variations/Prov. Sums	No preliminaries allowance where Provisional Sum work instructed is defined work.
13.5.5	Variations/Prov. Sums	Change of conditions under which other work executed

INSTRUCTIONS

Clause No	Clause Title	Signpost
13.5.6.6	Variations/Prov. Sums	Valuation of instructions issued with regard to Performance Specified Work
13A	Variation instruction	Application of cl.13A
13A.1.1	Variation instruction	Content of instruction issued under cl.13A
13A.1.1	Variation instruction	Procedure if insufficient information is provided in instruction
13A.1.2	Variation instruction	Submission and timing of 13A Quotation
13A.2	Variation instruction	13A Quotation to indicate resources and methods if required by the instruction
13A.2.1	Variation instruction	13A Quotation to include the effect of the instruction on any other work
13A.3.2.2	Variation instruction	Upon acceptance of 13A Quotation Architect to state adjustment of Contract Sum
13A.4	Variation instruction	Procedure if Employer does not accept 13A Quotation by expiry of period
13A.8	Variation instruction	Valuation of work instructed varying work the subject of an accepted 13A Quotation
17.2	Practical Compl./Defects	Schedule of Defects
17.2; 17.3	Practical Compl./Defects	Responsibility for cost of work in making good defects
17.3	Practical Compl./Defects	Timing of requirement for making good defects
22A.5.3	All risks - Contractor	Restoration of work damaged by act of terrorism treated as an instructed Variation
22B.3.5	All risks - Employer	Restoration of work damaged treated as if it were an instructed Variation
22B.4.3	All risks - Employer	Restoration of work damaged by act of terrorism treated as an instructed Variation
22C.4.4.2	Insurance - existing structures	Restoration of work damaged treated as if it were an instructed Variation

Clause No	Clause Title	Signpost
22C.5.3	Insurance - existing structures	Restoration of work damaged by act of terrorism treated as an instructed Variation
22D.1	Insurance: Loss of LADs	Contractor may be instructed to obtain quotations for insurance
22D.1	Insurance: Loss of LADs	Contractor may be instructed to accept quotations for insurance
22FC.3.1	Joint Fire Code	Carrying out of Remedial Measures in accordance with instructions
23.2	Possession, completion	Postponement
25.3.1.4	Extension of time	Regard given to instructions requiring an omission
25.3.2	Extension of time	Regard given to instructions requiring an omission
25.3.3.2	Extension of time	Regard given to instructions requiring an omission
25.4.5	Extension of time	Classes of instruction regarded as Relevant Events
26.1	Loss and expense	QS may be instructed to ascertain
26.2.1	Loss and expense	Non receipt within a reasonable time
26.2.5	Loss and expense	Postponement
26.2.7	Loss and expense	Variations or expenditure of Provisional Sums
26.4.1	Loss and expense	QS may be instructed to ascertain for NSCs
27.2.1.3	Determination by Employer	Non-compliance with Architect instruction as ground
28.2.2.2	Determination by Contractor	Suspension of the Works caused by instructions issued
28A.1.1.4	Determination by either	Suspension caused by negligence etc. of local authority or Statutory Undertaker

INSTRUCTIONS

Clause No	Clause Title	Signpost
30.2.1.5	Certificates/payments	Profit on nominated firms' work subject to Retention
30.4A.1	Certificates/payments	QS may have to prepare statement of Retention that would have been deducted
30.5.2.1	Certificates/payments	QS may have to prepare statement of Retention
30.6.1.1	Certificates/payments	Documents sent to QS for adjustment of Contract Sum
30.6.2.8	Certificates/payments	Nomination instruction in respect of Nominated Suppliers
30.6.2.9	Certificates/payments	Profit on nominated firms' work in Contract Sum adjustment
30.6.2.12	Certificates/payments	Work set against Provisional Sums
30.9.1.1	Certificates/payments	Final Certificate as conclusive evidence that quality of instructed work is accepted
31.12	CIS	By Inland Revenue concerning treatment of previous error in calculation
34.2	Antiquities	Procedure for dealing with an object discovered
34.3.1	Antiquities	Involvement in loss and expense

Clause No	Clause Title	Signpost
	Part 2	
35.1.2	General	Nomination in the expenditure of Provisional Sums
35.1.3	General	Nomination in the issue of instructions requiring a Variation
35.2.1	General	Position where the intention to nominate is included in cl.13.3 instructions
35.4	Procedure	Identification term for Standard Form of Nomination Instruction
35.5.1	Procedure	Contractor's objection to a nomination to be within 7 days of instruction
35.5.2	Procedure	Architect may issue further instructions to remove objection
35.5.2	Procedure	Architect may cancel nomination instruction
35.5.2	Procedure	Architect may omit the work that was subject to nomination
35.5.2	Procedure	Architect may nominate another Sub-Contractor
35.5.2	Procedure	Copies of any instruction under c.35.5.2 to be sent to the Sub-Contractor
35.6	Procedure	Nomination instruction to be issued on NSC/N with a copy to the Sub-Contractor
35.7	Procedure	Action by Contractor on receipt of nomination instruction
35.8	Procedure	Action by Contractor if he fails to comply with cl.35.7 within 10 days.
35.9.2	Procedure	Architect considers that matters in Contractor's notice justify non-compliance
35.9.2	Procedure	Architect may issue further instructions to enable Contractor to comply
35.9.2	Procedure	Architect may cancel nomination instruction

INSTRUCTIONS

Clause No	Clause Title	Signpost
35.9.2	Procedure	Architect may omit the work that was subject to nomination
35.9.2	Procedure	Architect may nominate another Sub-Contractor
35.13.3	Payment	Adjustments where pre-nomination payments have been paid to NSC.
35.18.1.1	Early Final Payment	Rectification of defects by a substituted Sub-Contractor
35.24.5	Re-nomination	Consequences of compliance with an instruction to re-execute compliant work
35.24.6.1	Re-nomination	Notifying NSC of defaults
35.24.6.1	Re-nomination	Need for further instruction before NSC employment can be determined
35.24.6.2	Re-nomination	Informing Architect whether NSC employment has been determined
35.25	Determination	Contractor may not determine employment of NSC without instruction from Architect
36.1.1.1	Nominated Suppliers	Supplier named for materials and goods
36.1.1.2	Nominated Suppliers	Expenditure of a Provisional Sum as Prime Cost Sum
36.1.1.2	Nominated Suppliers	Supplier named for materials and goods
36.1.1.3	Nominated Suppliers	Expenditure of a Provisional Sum where only one supplier
36.1.1.3	Nominated Suppliers	Expenditure of a Provisional Sum as Prime Cost Sum
36.1.1.4	Nominated Suppliers	Variation requiring materials or goods from sole supplier
36.2	Nominated Suppliers	Nominating a supplier

Clause No	Clause Title	Signpost
36.3.1	Nominated Suppliers	Definition of amounts "properly chargeable"
36.4.3	Nominated Suppliers	Include cls.13.2&3 instructions as grounds for Variation of delivery programme
Part 4		
41A.5.8	Adjudication	Dispute as to whether an instruction under cl.8.4.4 is reasonable
41A.5.8.1	Adjudication	Adjudicator to have experience relevant to the instruction
41A.5.8.2	Adjudication	Adjudicator to appoint an independent expert having relevant
41A.5.8.4	Adjudication	Adjudicator's instructions to be supplied to parties
Part 5		
42.3	Performance Specified Work	Content of Contractor's Statement
42.4	Performance Specified Work	Timing for provision of Contractor's Statement
42.8	Performance Specified Work	Expenditure of Provisional Sums other than those specifically for PSW
42.10	Performance Specified Work	Correction of error in Contract Bills treated as if a Variation required by an instruction
42.11	Performance Specified Work	Instructions for Variations to PSW
42.12	Performance Specified Work	Limitation to permissible scope of Variations to PSW

INSTRUCTIONS

Clause No	Clause Title	Signpost
42.14	Performance Specified Work	Instructions necessary for integration of PSW with main Works
42.15	Performance Specified Work	Injurious affection arising from compliance with an instruction
	Appendices	
2	Code of Practice	Criteria to be considered in issuing cl.8.4.4 instructions
2.8	Code of Practice	Failure to carry out instructed tests
VAT 3.1	VAT Agreement	Appeal against decision of the Commissioners
4(iii) (a)	Retention Bond	Demand to state costs incurred by failure by Contractor to comply with instructions

Clause No	Clause Title	Signpost
	Part 1	
20	Injury and indemnity	
21	Insurance: persons etc.	**Principal applicable clauses**
22	Insurance: Works (incl: 22A, 22B, 22C, 22D, 22FC)	
1.3	Definitions etc.	Reference to All Risks Insurance and Joint Names Policy
16.1	Mats. and goods unfixed	Insurance of materials and goods stored on-site
16.2	Mats. and goods unfixed	Insurance of materials and goods stored off-site
18.1	Partial possession	Identification of relevant part as referred to in cls. 22.3.1 and 22C.1
18.1.3	Partial possession	Reduction of value insured under cl.22A, B or C
23.3.1	Possession, completion	Contractor retains possession until issue of CPC for purposes of insurance
23.3.2	Possession, completion	If Employer occupies premises before issue of CPC, insurers must be notified
23.3.3	Possession, completion	Additional premium arising from cl.23.3.2 is added to the Contract Sum
30.2.1.1	Certificates/payments	Amounts in Interim Certificates subject to Retention
30.2.2	Certificates/payments	Amounts in Interim Certificates not subject to Retention
30.3.5	Certificates/payments	Proof of insurance of listed materials
30.6.2.10	Certificates/payments	C1.21.2.3 provisions in adjustment of Contract Sum

INSURANCE

Clause No	Clause Title	Signpost
30.6.2.14	Certificates/payments	Amounts paid by Contractor under cls.22B or 22C in adjustment of Contract Sum
Part 2		
35.19.2	Early Final Payment	Provisions of cls. 22A, 22B or 22C unaffected by a final payment to NSC
Part 3		
39.1.1	Fluctuations – costs, taxes etc	Contract Sum includes cost of employer's liability insurance
39.1.2	Fluctuations – costs, taxes etc	Cost of employer's liability insurance is adjustable
Appendices		
21.1.1	Appendix	Amount of cover
21.2.1	Appendix	Whether required and amount of indemnity
22.1	Appendix	Choice of appropriate clause
22A, -B.1, -C.2	Appendix	Percentage to cover professional fees
22A.3.1	Appendix	Annual renewal date
22D	Appendix	Whether insurance required for loss of LADs insurance

Clause No	Clause Title	Signpost
22D.2	Appendix	Period for loss of LADs insurance
22FC.1	Appendix	Whether Joint Fire Code applies and state whether project is classified as large
3(b)	Off-site Materials etc. Bond	Contractor has agreed to insure listed items
4(iii)(b)	Retention Bond	Insurance premiums paid by Employer where Contractor fails to insure, as a demand

LIQUIDATED DAMAGES

Clause No	Clause Title	Signpost
	Part 1	
22D;	Insurance: Loss of LADs	**Principal applicable clauses**
24	Damages	
18.1.4	Partial possession	Proportionate reduction when partial possession has occurred
	Appendices	
22D	Appendix	Whether insurance required for loss of LADs insurance
24.2	Appendix	Rate of liquidated damages
VAT 2. (1; 2; 3)	VAT Agreement	Disregarded in calculating amounts of VAT

Clause No	Clause Title	Signpost
	Part 1	
8	Materials, goods etc.	**Principal applicable clauses**
30.3	Certificates/payments	
1.3	Definitions etc.	Site materials
1.5	Definitions etc.	Contractor remains wholly responsible the Works including all materials or goods
2.1	Contractor's obligations	Quality and standards as specified in Contract Documents
6.1.4.1	Statutory obligations	Supply in an emergency
6.1.4.3	Statutory obligations	Conditions under which materials supplied under cl.6.1.4.1 will be a Variation
6.2.1	Statutory obligations	Conditions under which fees and charges will not be added to the Contract Sum
13.1.1.2	Variations/Prov. Sums	Alteration in kind and standard as part of the definition of a Variation
13.1.1.3	Variations/Prov. Sums	Removal of compliant materials and goods as part of the definition of a Variation
13.1.3	Variations/Prov. Sums	Nomination for measured work excluded from the definition of a Variation
13.5.4	Variations/Prov. Sums	Work evaluated on the basis of daywork
16.1	Mats. and goods unfixed	Unfixed materials may not be removed without consent
16.1	Mats. and goods unfixed	Status of materials when value included in an Interim Certificate
17.2	Practical Compl./Defects	Defects appearing in Defects Period are to be specified
17.3	Practical Compl./Defects	Defects at any time may be the subject of an instruction

MATERIALS/GOODS

Clause No	Clause Title	Signpost
19.4.2.1	Assignment/Sub-Contracts	Conditions, to be included in any Sub-Contract, relating to unfixed materials
19.4.2.2	Assignment/Sub-Contracts	Conditions, to be included in any Sub-Contract, relating to materials paid by Employer
19.4.2.3	Assignment/Sub-Contracts	Conditions, to be included in any Sub-Contract, relating to materials paid by Contractor
19.4.2.4	Assignment/Sub-Contracts	Cls.19.4.2.1 -.3 are without prejudice to property transfer rules of cl.30.3
19.5.1	Assignment/Sub-Contracts	Contractor is responsible for all materials including those supplied by NSC.
19.5.2	Assignment/Sub-Contracts	No requirement for Contractor to supply and fix for NSC work
22A.1	All risks - Contractor	Position on VAT relative to reinstatement value
22B.1	All risks - Employer	Position on VAT relative to reinstatement value
22C.2	Insurance - existing structures	Position on VAT relative to reinstatement value
25.4.4	Extension of time	Strikes affecting manufacture etc. are a Relevant Event
25.4.5.2	Extension of time	Testing may be a Relevant Event
25.4.8.2	Extension of time	Failure by Employer to supply may be a Relevant Event
25.4.9	Extension of time	Exercise of Statutory power may be a Relevant Event
25.4.10.2	Extension of time	Inability to secure may be a Relevant Event
26.2.2	Loss and expense	Opening for inspection may be a ground for claim
26.2.4.2	Loss and expense	Failure by Employer to supply may be a ground for claim
27.2.1.3	Determination by Employer	Contractor failure to comply with a written notice to remove non-compliant materials

Clause No	Clause Title	Signpost
27.6.1	Determination by Employer	Employer may use and/or purchase as necessary
27.6.2.1	Determination by Employer	Assignment of agreements if required
27.6.2.2	Determination by Employer	Employer may pay suppliers if the price is not already paid
27.6.3	Determination by Employer	Contractor to remove his property when so required
28.4	Determination by Contractor	Effect on accrued rights and on cl.20 liabilities of removal of materials etc.
28.4.1	Determination by Contractor	Timing of removal of materials
28.4.3.5	Determination by Contractor	Contractor's account to include for paid and/or contracted materials.
28.2.2.4	Determination by Contractor	Payment to Contractor for those properly ordered
28A.3	Determination by either	Timing of removal of materials
28A.5.4	Determination by either	Employer's account to include for Contractor's paid and/or contracted materials
30.2.1.2	Certificates/payments	Materials on site included in Interim Certificates and subject to Retention
30.2.1.3	Certificates/payments	Materials off site included in Interim Certificates and subject to Retention
30.2.2.1	Certificates/payments	Amounts incurred under cl.8.3 not subject to Retention
30.2.3.1	Certificates/payments	Amounts deductible under cl.8.4.2 not subject to Retention
30.3	Certificates/payments	Definition of 'the listed items'.
30.3	Certificates/payments	Whole clause deals with the treatment of 'listed items'

MATERIALS/GOODS

Clause No	Clause Title	Signpost
30.4.1.2	Certificates/payments	Retention which may be deducted
30.4.2	Certificates/payments	Retention on payments to Contractor and NSCs to be separately identified
30.6.2.4	Certificates/payments	Amounts deductible under cl.8.4.2 in adjustment of Contract Sum
30.6.2.8	Certificates/payments	Payment for supplies by Nominated Suppliers in adjustment of Contract Sum
30.6.2.10	Certificates/payments	Amounts incurred under cl.8.3 in adjustment of Contract Sum
30.9.1.1	Certificates/payments	Final Certificate as conclusive evidence of quality etc
30.9.1.2	Certificates/payments	Final Certificate as conclusive evidence of accuracy
30.10	Certificates/payments	Effect of certificates other than Final Certificate
31.1	CIS	Definition of 'the direct cost of materials'
31.5.1.1	CIS	Under a CIS4, Contractor to give a statement showing direct cost of materials
31.5.3	CIS	Employer to make estimate of cost of materials if Contractor fails under cl.31.5.1.1
	Part 2	
36	Nominated Suppliers	**Principal applicable clauses**
35.1	General	Reservation of right to nominate
35.1.3.2	General	Instructions for further NSC work or goods of similar type
35.21.2	Clause 2.1 of NSC/W	Contractor not responsible for selection of materials

Clause No	Clause Title	Signpost
35.21	Clause 2.1 of NSC/W	Nothing in cl.35.21 affects obligations of Contractor
35.21.4	Clause 2.1 of NSC/W	Loss or damage before Practical Completion of Works
35.24.5	Re-nomination	Re-execution of work as a result of an instruction or other power of Architect
35.24.6.3	Re-nomination	Further nomination of a Sub-Contractor where cl. 35.24.1 applies
35.24.7.3	Re-nomination	Further nomination of a Sub-Contractor where cl. 35.24.2 applies
35.24.7.4	Re-nomination	Further nomination of a Sub-Contractor where cl. 35.24.4 applies
35.24.8.1	Re-nomination	Further nomination of a Sub-Contractor where cl. 35.24.3 applies
35.26.1	Determination	Architect to direct Contractor as to amounts of loss and expenses incurred by him
35.26.2	Determination	Architect to issue Interim Certificate with value of materials/goods supplied by NSC
Part 3		
38.1.2.2	Fluctuations – taxes etc.	Calculation of Contract Sum and effect of changes to tender type or rate
38.2.1	Fluctuations – taxes etc.	Calculation of Contract Sum
38.2.2	Fluctuations – taxes etc.	Effect of changes to tender type or rate
38.5.2	Clause 38 provisions	Materials and goods to which cls.38.1 to .3 do not apply
38.6.2	Clause 38 provisions	Definition of materials and goods
39.1.1.2	Fluctuations – costs, taxes etc	Calculation of Contract Sum and effect of changes to tender type or rate

MATERIALS/GOODS

Clause No	Clause Title	Signpost
	Part 1	
2.2.2.1	Contractor's obligations	Standard Method of Measurement 7th Edition (SMM7)
2.2.2.2	Contractor's obligations	Procedure where defined work is not as required by SMM7
13.1.3	Variations /Prov. Sum	Nomination where measured quantities appear in the Bills is not a Variation
13.5.1	Variations /Prov. Sum	Valuation of work which can be properly measured
13.5.3.1	Variations /Prov. Sum	In valuations, measurement to be in accordance with principles used for Contract Bills
13.5.3.3; .6.4	Variations /Prov. Sum	Adjustment of preliminaries of the type referred to in SMM7
13.5.4	Variations /Prov. Sum	Valuation of work which cannot be properly measured
13.6	Variations /Prov. Sum	Contractor may be present when QS takes measurements
19.3.1	Assignment/Sub-Contracts	Provisions for measured work carried out by named firm
	Part 2	
35.1.4	General	Clause 35.1 applies notwithstanding Rule A51 of SMM7
	Part 4	
41B.2	Arbitration	Arbitrator has power to direct measurements to be made

NOMINATED SUB-CONTRACTORS

Clause No	Clause Title	Signpost
	Part 1	
1.3	Definitions etc	3.3A Quotation
1.3	Definitions etc	Nominated Sub-Contract
1.3	Definitions etc	Nominated Sub-Contractor
6.2.1	Statutory obligations	Fees and charges where Local Authority etc. working as NSC
6.3	Statutory obligations	Local Authority etc. working in pursuance of Statutory duty
8.4.2; .3	Work, materials and goods	Contractor to consult with relevant NSC when work or materials are non-compliant
8.5	Work, materials and goods	Contractor to consult with relevant NSC where there is failure to comply with cl.8.1.3
11	Access to the Works	Access to the workshops of NSC
13.1.3	Variations/Prov. Sums	Nomination for work measured in Contract Bills
13.3.2	Variations/Prov. Sums	Instruction for the expenditure of Provisional Sums in a Nominated Sub-Contract
13.4.1.3	Variations/Prov. Sums	Valuation of Variations in accordance with NSC/C
13.4.2	Variations/Prov. Sums	Valuation of work by Contractor for which he has tendered under cl.35.2
13A.1.1	Variation Instruction	Instruction for a 13A Quotation to include information for any NSC 3.3A Quotation
13A.1.2	Variation Instruction	Timing for submission of 13A with 3.3A Quotation
13A.1.3	Variation Instruction	Variation covered by 13A with 3.3A Quotation not to be carried out until accepted
13A.2.1	Variation Instruction	13A Quotation to include for any effect the work may have on that of any NSC

Clause No	Clause Title	Signpost
13A.3.2.3	Variation Instruction	An acceptance is to include for any revisions to Completion Date of NSC work
13A.3.2.4	Variation Instruction	An acceptance is to include any 3.3A Quotation given as part of the tender
13A.7	Variation Instruction	Contractor to notify NSC of any change in time for submission of 13A Quotation
19.2.1	Assignment/Sub-Contracts	NSC is not a Domestic Sub-Contractor
19.5.1	Assignment/Sub-Contracts	Provisions relating to NSCs are set out in PART 2
19.5.2	Assignment/Sub-Contracts	Contractor not required to perform duties of NSC
22.3.1	Insurance of the Works	NSCs to be included in Joint Names Policy
22A.4.4	All Risks – by Contractor	Authority for insurance monies to be paid to the Employer
22B.3.4	All Risks – by Employer	Authority for insurance monies to be paid to the Employer
22C.1	Insurance - existing structures	Authority for insurance monies to be paid to the Employer
22C.4.2	Insurance - existing structures	Authority for insurance monies to be paid to the Employer
25.2.1.2	Extension of time	Copies to NSCs of notice defining causes of delay
25.2.2.2	Extension of time	Copies to NSCs of notice estimating extent and effect of delay
25.2.3	Extension of time	Copies to NSCs of notice updating previous particulars
25.3.5	Extension of time	Notification of decisions fixing Completion Date
25.4.5.1	Extension of time	Compliance with cl.35 instructions is a Relevant Event
25.4.7	Extension of time	Delay by NSC is a Relevant Event

NOMINATED SUB-CONTRACTORS

Clause No	Clause Title	Signpost
26.4.1	Loss and expense	Applications in writing under cl.4.38.1 of NSC/C
26.4.2	Loss and expense	Copies to NSC of notice of revised Sub-Contract Completion Date
28.2.2.2	Determination by Contractor	Suspension of the works by reason of architect's instructions as a ground
28.4.3	Determination by Contractor	Relevant amounts in respects of NSCs to be included in Contractor's account
28A.5	Determination by either	Contractor to provide Employer with NSC documents for preparation of an account
28A.7	Determination by either	Employer to advise on proportion of payments attributable to NSCs
30.1.2.2	Certificates/payments	Applications for Interim payments to include applications by NSCs
30.1.2.2	Certificates/payments	Quantity Surveyor to give detail of extent of disagreement with NSC application
30.2.1.4	Certificates/payments	Amounts in Interim Certificates subject to Retention
30.2.2.3	Certificates/payments	Amounts in Interim Certificates not subject to Retention
30.2.2.5	Certificates/payments	Amounts in Interim Certificates not subject to Retention
30.2.3.2	Certificates/payments	Amounts deducted from Interim Certificates not subject to Retention
30.4.1.2	Certificates/payments	Amounts against which full Retention may be held
30.4.1.3	Certificates/payments	Amounts against which half Retention may be held
30.4.2	Certificates/payments	Definition of Nominated Sub-Contract Retention
30.4A.1	Certificates/payments	Architect or QS to prepare statement of what Retention would have been (Bond)
30.4A.3	Certificates/payments	Employer shall release Retention held before production of Retention Bond

Clause No	Clause Title	Signpost
30.5.1	Certificates/payments	Interest in Retention fiduciary as trustee for NSC
30.5.2	Certificates/payments	Copy of Statement of Retention to be sent to NSC
30.5.2.1	Certificates/payments	Statement of Retention held against NSCs to be prepared
30.5.2.2	Certificates/payments	NSCs to be sent copy of Statement of Retention where relevant
30.5.3	Certificates/payments	NSC may request that Retention be placed in a separate account
30.5.4	Certificates/payments	Deduction against NSC Retention to be notified to Contractor
30.6.1.1	Certificates/payments	Documents necessary for adjustment of Contract Sum
30.6.1.2	Certificates/payments	Copy to each NSC of statement of final Valuations
30.6.2.6	Certificates/payments	Final value of NSCs in adjustment of Contract Sum
30.6.2.7	Certificates/payments	Tender or adjusted tender values under cl.35.2 in adjustment of Contract Sum
30.7	Certificates/payments	A Certificate to include all final amounts due to NSCs
30.8	Certificates/payments	Architect to inform NSCs of date of Final Certificate issue
	Part 2	
35	NSC Rules	**Principal applicable clause**

NOMINATED SUB-CONTRACTORS

Clause No	Clause Title	Signpost
	Part 3	
38.5.2	Clause 38 provisions	Not applicable to work done by NSC
38.5.3	Clause 38 provisions	Not applicable to tender or adjusted tender values under cl.35.2
39.6.2	Clause 39 provisions	Not applicable to work done by NSC
39.6.3	Clause 39 provisions	Not applicable to tender or adjusted tender values under cl.35.2
	Part 5	
42.18	Performance Specified Work	Nominated Sub-Contractors shall not provide PSW
	Appendices	
35.2	Appendix	Work reserved for NSCs for which contractor desires to tender

Clause No	Clause Title	Signpost
	Part 1	
1.3	Definitions etc.	Nominated Supplier
6.2.1	Statutory obligations	Fees and charges where Local Authority etc. acting as Nominated Suppliers
25.4.5.1	Extension of time	Compliance with cl.36 instructions is a Relevant Event
25.4.7	Extension of time	Delay by Nominated Suppliers is a Relevant Event
28A.5	Determination by either	Contractor to provide Nominated Suppliers documents for preparation of accounts
30.6.1.1	Certificates/payments	Documents necessary for adjustment of Contract Sum
30.6.2.8	Certificates/payments	Final value of Nominated Suppliers in adjustment of Contract Sum
	Part 2	
36.1 - 5	Nominated Suppliers	**Principal applicable clauses**
	Part 3	
38.5.2	Clause 38 provisions	Not applicable to goods supplied by Nominated Suppliers
39.6.2	Clause 39 provisions	Not applicable to goods supplied by Nominated Suppliers
	Part 5	
42.18	Performance Specified Work	Nominated Suppliers shall not provide PSW

NOTICE/NOTIFY

Clause No	Clause Title	Signpost
	Part 1	
1.6	Definitions etc.	Employer notifies Contractor of new Planning Supervisor
1.7	Definitions etc.	Manner of serving notices
1.9	Definitions etc.	Employer notifies Contractor of Employer's Representative
2.3.5	Contractor's obligations	Contractor notifies Architect of discrepancies between documents
2.4.1	Contractor's obligations	Contractor notifies Architect of discrepancies between Statement of PSW and other
4.1.2	Architect's instructions	Employer's action following non compliance with written instruction
6.1.1	Statutory obligations	Contractor to comply with all Statutory notices
6.1.2	Statutory obligations	Divergence between documents and Statutory requirements
6.1.3	Statutory obligations	Instructions to be given within 7 days of receipt of notice
6.1.6	Statutory obligations	Notice given by either party where divergence found with Contractor's Statement
6A.2	CDM Regulations	Contractor notifies Employer of any change to H&S Plan
6A.2	CDM Regulations	Employer notifies Planning Supervisor and Architect of any change to H&S Plan
13.4.1.2.A2	Contractor's Price Statement	QS notifies Contractor of acceptance or otherwise of Price Statement
13.4.1.2.A4.1	Contractor's Price Statement	QS notifies Contractor of reasons for non-acceptance
13.4.1.2.A5	Contractor's Price Statement	Position in the absence of notification under cl. A2
13.4.1.2.A7.1	Contractor's Price Statement	QS notifies Contractor with regard to requirements under paras. A1.1/2

Clause No	Clause Title	Signpost
13.4.1.2.A7.2	Contractor's Price Statement	Position in the absence of notification under cl.A7.1
13A.3.1	Variation instruction	Employer notifies Contractor if he wishes to accept 13A Quotation
13A.7	Variation instruction	Contractor notifies NCSs of any change in timing for issue of 13A Quotation
18.1.4	Partial Possession	Employer notifies Contractor if he will deduct sums from monies due
22A.4.1	All Risks – by Contractor	Contractor notifies Architect and Employer of Joint Names Policy losses
22A.5.1	All Risks – by Contractor	Procedure if insurers notify Contractor or Employer of termination of terrorism cover
22A.5.2	All Risks – by Contractor	Employer notification to Contractor; action upon receipt of notice under cl.22A.5.1
22A.5.2.2	All Risks – by Contractor	Notice under cl.22A.5.2 to give date of any termination of employment
22B.3.1	All Risks – by Employer	Contractor notifies Architect and Employer of Joint Names Policy losses
22B.4.1	All Risks – by Employer	Procedure if insurers notify Contractor or Employer of termination of terrorism cover
22B.4.2	All Risks – by Employer	Employer notification to Contractor; action upon receipt of notice under cl.22B.4.1
22B.4.2.2	All Risks – by Employer	Notice under cl.22B.4.2 to give date of any termination of employment
22C.1A.1	Insurance - existing structures	Procedure if insurers notify Contractor or Employer of termination of terrorism cover
22C.1A.1	Insurance - existing structures	Employer notification to Contractor; action upon receipt of notice under cl.22C.1A.1
22C.1A.1.2	Insurance - existing structures	Notice under cl.22C.1A.1 to give date of any termination of employment
22C.4	Insurance - existing structures	Contractor notifies Architect and Employer of Joint Names Policy losses
22C.4.3.1	Insurance - existing structures	Notice of determination by either party

NOTICE/NOTIFY

Clause No	Clause Title	Signpost
22C.4.3.1	Insurance - existing structures	Procedure if notice served under cl.22C4.3.1
22C.4.3.2	Insurance - existing structures	Application of cls.28A.4/5 if notice served under cl.22C.4.3.1
22C.4.4	Insurance - existing structures	Procedure if no notice served or notice under cl.22C.4.3.1 is not upheld
22C.5.1	Insurance - existing structures	Procedure if insurers notify Contractor or Employer of termination of terrorism cover
22C.5.2	Insurance - existing structures	Employer notification to Contractor; action upon receipt of notice under cl.22C.5.1
22C.5.2.2	Insurance - existing structures	Notice under cl.22C.5.2 to give date of any termination of employment
22FC.3.1	Joint Fire Code	Specification of Remedial Measures
22FC.3.2	Joint Fire Code	Procedure if Contractor does not put Remedial Measures in hand
23.3.2	Possession, completion	Contractor or Employer notify insurers if Employer occupies part of site or works
23.3.3	Possession, completion	Contractor notifies Employer of any additional premiums where cls. 22A2 or .3 apply
24.2.1.2	Damages	Employer notifies Contractor of liquidated damages
25.1	Extension of time	Notice shall include further notice
25.2.1.1	Extension of time	Circumstances causing delay
25.2.1.2	Extension of time	Copies to NSCs
25.2.2	Extension of time	Particulars of effect of Relevant Event
25.2.2.2	Extension of time	Particulars in cl.25.2.2 to be given to NSC
25.2.3	Extension of time	Further written notices

Clause No	Clause Title	Signpost
25.3.1	Extension of time	Action by Architect upon receipt of notice under cls.25.2.1.1, 25.2.2 and .3
25.3.1	Extension of time	Decision to fix new Completion Date
25.3.1	Extension of time	Decision not to fix new Completion Date
25.3.5	Extension of time	Every NSC to be notified of decisions on Completion Date
25.4.12	Extension of time	Failure to give access to site after receipt of notice
26.2.6	Loss and expense	Failure to give access to site after receipt of notice
27.1	Determination by Employer	To be in writing
27.1	Determination by Employer	Rules concerning delivery
27.2.1	Determination by Employer	Architect notifies Contractor of default
27.2.1.3	Determination by Employer	Refusal to comply with written notice
27.2.2	Determination by Employer	Procedure if Contractor continues a specified default
27.2.3	Determination by Employer	Procedure if Contractor repeats a specified default
27.2.4	Determination by Employer	Determination notices not to be given unreasonably
27.3.4	Determination by Employer	Employer notifies Contractor of determination
27.5.1	Determination by Employer	Employer is not bound by payment provisions from date when notice could be given
27.6.4.1	Determination by Employer	Contractor's rights to payment related to date when notice could be given
27.7.1	Determination by Employer	Employer's decision not to have the Works completed

NOTICE/NOTIFY

Clause No	Clause Title	Signpost
27.7.1	Determination by Employer	Statement of account sent with reasonable time of 27.7.1 notification
27.7.2	Determination by Employer	Requiring Employer to state whether cls.27.6.1-6 are to apply
28.1	Determination by Contractor	To be in writing
28.1	Determination by Contractor	Rules concerning delivery
28.2.1	Determination by Contractor	Contractor notifies Architect of default
28.2.2.4	Determination by Contractor	Failure to give access to site after receipt of notice as an event
28.2.2	Determination by Contractor	Contractor notifies Architect of events
28.2.3	Determination by Contractor	Procedure if Employer continues a specified default or event
28.2.4	Determination by Contractor	Procedure if Employer repeats a specified default or event
28.2.5	Determination by Contractor	Determination notices not to be given unreasonably
28.3.3	Determination by Contractor	Contractor notifies Employer of determination
28A.1.1	Determination by either	Contractor notifies Employer or Employer notifies Contractor of default
28A.1.1	Determination by either	Rules concerning delivery
28A.1.2	Determination by either	Contractor may not give notice in the event of his own negligence
30.1.1.3	Certificates/payments	Employer notifies Contractor of proposed payment
30.1.1.4	Certificates/payments	Employer notifies Contractor of amounts to be withheld
30.1.1.5	Certificates/payments	Payment in the absence of notification under cls.30.1.1.3/4

Clause No	Clause Title	Signpost
30.1.4	Certificates/payments	Right of suspension of Contractor's obligations
30.8.1	Certificates/payments	Calculation of Final Certificate taking account of amounts withheld under cl.30.1.1.4
30.8.2	Certificates/payments	Employer notifies Contractor of amount proposed to be paid
30.8.3	Certificates/payments	Employer notifies Contractor of amounts to be withheld from final balance due
30.8.4	Certificates/payments	Payment in the absence of notification under cls.30.8.2/3
31.4.1	CIS	Employer notifies Contractor of dissatisfaction with validity of Authorisation
31.4.2	CIS	Procedure following notification under cl.31.4.1
31.8	CIS	Contractor notifies Employer of any withdrawal by inland Revenue of CIS 5 or 6
	Part 2	
35.8	Procedure	Contractor notifies Architect of non-compliance with cl.35.7
35.8.2	Procedure	Contractor identifies reasons for non-compliance with cl.35.7
35.9	Procedure	Action of Architect upon receipt of notice under cl.35.8
35.9.2	Procedure	Architect comment on content of notice under cl.35.8
35.14.2	Extension of periods	Action of Architect upon receipt of notice requesting extension of period
35.15.1	Failure to complete	Contractor notifies Architect of failure by NSC to complete
35.15.2	Failure to complete	Timing of certificate issued under cl.35.15.1

NOTICE/NOTIFY

Clause No	Clause Title	Signpost
35.24.6.1	Re-nomination	Instruction by Architect to Contractor to give notice specifying default by NSC
35.24.6.2	Re-nomination	Contractor informs Architect if determination has followed the cl.35.24.6.1 notice
	Part 3	
38.4.1	Clause 38 provisions	Contractor notifies Architect of the occurrence of events
38.4.2	Clause 38 provisions	Timing of notice under cl.38.4.1
38.4.2	Clause 38 provisions	Written notice is a condition precedent to payment
38.4.8.2	Clause 38 provisions	Cl.38.4.7 applies if Architect fixes Completion Date in response to notice under cl.25
39.5.1	Clause 39 provisions	Contractor notifies Architect of the occurrence of events
39.5.2	Clause 39 provisions	Timing of notice under cl.39.5.1
39.5.2	Clause 39 provisions	Written notice is a condition precedent to payment
39.5.8.2	Clause 39 provisions	Cl.39.5.7 applies if Architect fixes Completion Date in response to notice under cl.25
40.7.2.2	Price adjustment formulae	Cl.40.7.1 applies if Architect fixes Completion Date in response to notice under cl.25
	Part 4	
41A.2.2	Adjudication	Appointment of Adjudicator where either party has given notice to refer a dispute
41A.2.2	Adjudication	Timing of application to nominator in relation to issue date of notice

Clause No	Clause Title	Signpost
41A.4.1	Adjudication	Procedure where either party has given notice to refer a dispute
41A.4.1	Adjudication	Timing for referral of dispute to Adjudicator
41A.5.3	Adjudication	Extension of period for reaching decision
41A.5.5.2	Adjudication	Adjudicator has power to review notices
41A.5.5.3	Adjudication	Adjudicator may demand further information than included in notice of referral
41A.5.5.6	Adjudication	Prior notice to parties if Adjudicator seeks further information from employees
41A.5.5.7	Adjudication	Prior notice to parties if Adjudicator seeks further information from others
41B.1.1	Arbitration	Service of notice of Arbitration
41B.1.3	Arbitration	Service of further notices of Arbitration
41B.2	Arbitration	Arbitrator has power to review notices
41B.4	Arbitration	Notice to apply to the courts
41B.6	Arbitration	Joint notice to Arbitrator to have proceedings conducted to amended JCT Rules

Part 5

42.5	Performance Specified Work	Architect notifies Contractor of requirement to amend Statement
42.6	Performance Specified Work	Architect notifies Contractor of any deficiency in Statement
42.15	Performance Specified Work	Contractor notifies Architect of any injurious affection arising from instructions

NOTICE/NOTIFY

Clause No	Clause Title	Signpost
	Appendices	
3(b)	Advance Payment Bond	Employer provides Surety with completed notice of demand
5	Advance Payment Bond	Actions that may be done without notice to the Surety
	Advance Payment Bond	Notice of Demand
4	Off-Site Materials Bond	Demand to be issued on a Notice of Demand
7	Off-Site Materials Bond	Actions that may be done without notice to the Surety
	Off-Site Materials Bond	Notice of Demand
VAT 1A.1	VAT Agreement	Application of cls.1.1 to 1.2.2 dependent upon notices under Cls. 1A.4 or 1A.2
VAT 1A.2	VAT Agreement	Contractor notifies Employer of applicable rates
VAT 1A.2	VAT Agreement	Timing for notification of Variations to rates
VAT 1A.3	VAT Agreement	Rates given on notice to be shown on Interim Certificates
VAT 1A.4	VAT Agreement	Contractor notifies Employer or Employer notifies Contractor if cl.1A ceases to apply
VAT 1.2.2	VAT Agreement	Reply to notification of objection
VAT 1.2.2	VAT Agreement	Objection to provisional assessment
VAT 1.3.4	VAT Agreement	Total tax under cl.1.3.3 exceeded by final statement
2	Retention Bond	Employer notifies Surety of date of next Interim Certificate after Practical Completion
4(iv)(2)	Retention Bond	Demand by Employer to certify Contractor has been given 14 days notice of liability

Clause No	Clause Title	Signpost
	Part 1	
1.3	Definitions etc.	Numbered Documents
2.3.5	Contractor's obligations	Discrepancy between Numbered Documents and other documents
13.4.1.3	Variations/Prov. Sums	Valuation of NSC work
26.2.3	Loss and expense	Discrepancy between Numbered Documents, Contract Drawings and Contract Bills
30.9.1.1	Certificates/payments	Effect of Final Certificate

PAYMENT

Clause No	Clause Title	Signpost
16.2	Mats. and goods unfixed	Materials and goods paid for but off-site
18.1.1	Partial possession	Date of Practical Completion of relevant part
18.1.4	Partial possession	Proportionate adjustment of liquidated damages
19.4.2.2	Assignment/Sub-Contracts	Upon payment of Contractor, property in Sub-Contract supplied materials passes
19.4.2.3	Assignment/Sub-Contracts	Upon payment by Contractor, property in Sub-Contract supplied materials passes
19.4.3	Assignment/Sub-Contracts	Sub-Contract to provide for payment of interest in event of late payment
21.1.3	Insurance: persons etc.	Amounts paid by the Employer may be deducted
21.2.1.9	Insurance: persons etc.	Sums payable as damages for breach of contract excluded from insurance
22A.2	All Risks – by Contractor	Amounts paid by the Employer may be deducted
22A.4.2	All Risks – by Contractor	Loss or damage disregarded in calculating amounts payable under the Contract
22A.4.4	All Risks – by Contractor	Authorisation to pay insurance monies to Employer
22A.4.4	All Risks – by Contractor	Insurance monies paid to Contractor by instalments
22A.4.5	All Risks – by Contractor	Contractor is not entitled to restoration money greater than that from insurance
22A.5.2.2	All Risks – by Contractor	Rules of payment change if terrorism cover withdrawn and employment determined
22A.5.3	All Risks – by Contractor	Restoration work treated as a Variation if instructed though terrorism cover withdrawn
22A.5.3	All Risks – by Contractor	No reduction in amounts payable to the Contractor for any alleged contributory action
22A.5.4.1	All Risks – by Contractor	Adjustment of Contract Sum if higher terrorism premium becomes payable.

PAYMENT

Clause No	Clause Title	Signpost
22B.2	All Risks – by Employer	Amounts paid by the Contractor may be added
22B.3.2	All Risks – by Employer	Loss or damage disregarded in calculating amounts payable under the Contract
22B.3.4	All Risks – by Employer	Authorisation to pay insurance monies to Employer
22B.4.2.2	All Risks – by Employer	Rules of payment change if terrorism cover withdrawn and employment determined
22B.4.3	All Risks – by Employer	Restoration work treated as a Variation if instructed though terrorism cover withdrawn
22B.4.3	All Risks – by Employer	No reduction in amounts payable to the Contractor for any alleged contributory action
22C.1	Insurance - existing structures	Authorisation to pay insurance monies to Employer
22C.1A.3	Insurance - existing structures	Rules of payment change if terrorism cover withdrawn and employment determined
22C.3	Insurance - existing structures	Amounts paid by the Contractor may be added
22C.4.1	Insurance - existing structures	Loss or damage disregarded in calculating amounts payable under the Contract
22C.4.2	Insurance - existing structures	Authorisation to pay insurance monies to Employer
22C.5.2.2	Insurance - existing structures	Rules of payment change if terrorism cover withdrawn and employment determined
22C.5.2.3	Insurance - existing structures	Restoration work treated as a Variation if instructed though terrorism cover withdrawn
22C.5.2.3	Insurance - existing structures	No reduction in amounts payable to the Contractor for any alleged contributory action
22D.1	Insurance for loss of LADs	Insurance to pay Employer for lost liquidated damages when cl.25 4.3 extension given
22D.3	Insurance for loss of LADs	Method of calculating payment
22FC.3.2	Joint Fire Code	Employer may employ and pay others to carry out Remedial Measures

Clause No	Clause Title	Signpost
24.2.1	Damages	Payment of liquidated damages
24.2.1.2	Damages	Employer may give notice of deduction for liquidated damages
24.2.2	Damages	Repayment of amounts when later Completion Date fixed
25.4.18	Extension of time	Delay arising from suspension by Contractor pursuant to cl.30.1.4 is Relevant Event
26.1	Loss and expense	Loss and expense where reimbursement not otherwise paid
26.2.10	Loss and expense	Delay arising from suspension by Contractor pursuant to cl.30.1.4 is a matter
27.5.1	Determination by Employer	Circumstances under cl. 27.3.4 when Employer is not bound to make further payment
27.5.3	Determination by Employer	Set-off may not operate on payments made under interim arrangements
27.6.1	Determination by Employer	Employment and payment of others to complete
27.6.2.1	Determination by Employer	Assignment without payment of benefit of agreements
27.6.2.2	Determination by Employer	Payment to Sub-Contractor or supplier where price not already paid
27.6.2.2	Determination by Employer	Payment to Sub-Contractor or supplier deducted from monies due
27.6.4.1	Determination by Employer	Without prejudicing other rights, provisions requiring further payment do not apply
27.6.5.2	Determination by Employer	Payment made to Contractor taken into account
27.6.5.3	Determination by Employer	Total amount that would have been payable, taken into account
27.6.6	Determination by Employer	Calculation of debt payable to either party
27.7.1	Determination by Employer	Calculation of debt payable to either party where Employer decides not to proceed

PAYMENT

Clause No	Clause Title	Signpost
28.2.1.1	Determination by Contractor	Failure to pay on a certificate
28.4	Determination by Contractor	Without prejudicing other rights, provisions requiring further payment do not apply
28.4.2	Determination by Contractor	Payment of Retention following determination
28.4.3	Determination by Contractor	Employer to pay amounts properly due within 28 days of account submission
28.4.3.5	Determination by Contractor	Contractor's account to include for payment made or due for materials or goods
28.4.3.5	Determination by Contractor	Upon payment if full, materials or goods become property of the Employer
28A.2	Determination by either	Upon determination, provisions requiring further payment do not apply
28A.4	Determination by either	Payment of Retention
28A.5	Determination by either	Employer to pay amounts properly due within 28 days of account submission
28A.5.4	Determination by either	Upon payment if full, materials or goods become property of the Employer
28A.7	Determination by either	NSCs to be informed of amounts attributable to them included in cl.28A.5 account
31.3	CIS	Employer to make no payment unless Authorisation has been provided
31.4.2	CIS	Authorisation must be valid
31.5.1	CIS	In the case of CIS4, procedures 7 days before payment is due
31.5.1.1	CIS	Contractor gives Employer a statement showing the direct cost of materials
31.5.1.2	CIS	Statutory deduction from payment not in respect of the direct cost of materials
31.6	CIS	No deductions where Authorisations are CIS5 or CIS6

PAYMENT

Clause No	Clause Title	Signpost
36.4.6	Nominated Suppliers	Timing for effecting full discharge for materials and goods
36.4.7	Nominated Suppliers	Passing of ownership whether or not payment made in full
Part 3		
40.2	Price adjustment formulae	Amendment to clause 30 for timing of interim valuations
40.6.1	Price adjustment formulae	Procedure if publication of Monthly Bulletins stops
40.6.2	Price adjustment formulae	Procedure if publication of Monthly Bulletins recommences
Part 4		
41A.5.5.2	Adjudication	Powers to open up etc. subject to cl.30.9
41A.5.5.8	Adjudication	Adjudicator may have regard Contract terms relating to the payment of interest
41A.5.7	Adjudication	Adjudicator may direct on payment of cost of tests or opening up.
41A.5.8.3	Adjudication	Adjudicator may direct on payment of expert fees.
41A.6.1	Adjudication	Adjudicator shall state how payment of his fee is to be apportioned
41B.2	Arbitration	Powers to rectify contract subject to cl.30.9

Clause No	Clause Title	Signpost
	Appendices	
30.1.1.6	Appendix	Advance payment
30.1.1.6	Appendix	Advance payment bond
30.1.3	Appendix	Dates of issue of Interim Certificates
30.2.1.1	Appendix	Listed items
30.3.1	Appendix	Payment of uniquely listed items
30.3.2	Appendix	Payment of items not uniquely listed
30.4.1.1	Appendix	Retention percentage
30.4A	Appendix	Retention Bond
	Advance Payment Bond	Entire document
6	Off-Site Materials Bond	Payments are made notwithstanding disputes
6	Off-Site Materials Bond	Payments are deemed valid for all purposes of the bond
Notice	Off-Site Materials Bond	Notice of payment demand
VAT 1	VAT Agreement	Supplies under a contract providing for periodic payment
VAT 1A.3	VAT Agreement	Payment made within the period for payment of certificates
VAT 1.2.1	VAT Agreement	VAT to be paid within period for payment of certificates
VAT 1.3.3	VAT Agreement	Timing for payment of balance of VAT

PAYMENT

Clause No	Clause Title	Signpost
VAT 1.3.4	VAT Agreement	Refund of overpayment
VAT 1.4	VAT Agreement	Receipt issued upon receipt of payments
VAT 2.1	VAT Agreement	Deductions for liquidated damages disregarded
VAT 2.3	VAT Agreement	Employer to pay the tax where cl.1A operates notwithstanding any deductions made.
VAT 3.1	VAT Agreement	Decisions of Commissioners before payment due
VAT 3.2	VAT Agreement	Payment where necessary before appeal can proceed
VAT 3.3	VAT Agreement	Timing for payment or refund after final adjudication
VAT 4	VAT Agreement	Conditions for discharge of Employer from further liability
VAT 4	VAT Agreement	Payment arising out of correction of error
VAT 5	VAT Agreement	VAT Agreement applicable to awards of an Arbitrator
VAT 7	VAT Agreement	Conditions under which Employer is not obliged to pay further
VAT 8	VAT Agreement	VAT arising out of determination under cl.27.4
4(iii)	Retention Bond	Statement of grounds for demand for payment of the bond

Clause No	Clause Title	Signpost
	Part 1	
13.5.6	Variations/Prov. Sums	**Principal applicable clause**
1.3	Definitions etc.	Performance Specified Work
1.3	Definitions etc.	Provisional Sum
2.4.1	Contractor's obligations	Divergence between Contractor's Statement and any subsequent instruction
5.9	Contract documents	Provision of as-built information
5.9	Contract documents	Provision of maintenance and operational information
6.1.7	Statutory obligations	Effect of change in Statutory Requirements on PSW
8.1.1	Work, materials and goods	Kinds and standards of materials and goods used
8.1.2	Work, materials and goods	Standards of workmanship
8.1.4	Work, materials and goods	No substitution of goods and materials for any described in Contractor's Statement
17.1	Practical Compl./Defects	Issue of CPC dependent upon compliance with cl.5.9
25.3.1.4	Extension of time	Architect shall state extent new Completion Date is affected by reduction in PSW
25.3.2	Extension of time	Fixing of an earlier Completion Date
25.3.3.2	Extension of time	Fixing of an earlier Completion Date
25.4.5.1	Extension of time	Instruction for expenditure of Provisional Sum for PSW is not a Relevant Event
25.4.15	Extension of time	Delay caused by change in Statutory Requirements affecting PSW is a Relevant Event
26.2.7	Loss and expense	Instruction for expenditure of Provisional Sum for PSW not a matter for loss/expense

PERFORMANCE SPECIFIED WORK

Clause No	Clause Title	Signpost
30.10.2	Certificates/payments	Certificates other than Final Certificate are not evidence of compliance.
	Appendices	
42	Performance Specified Work	**Principal applicable clause**

POSTPONEMENT

Clause No	Clause Title	Signpost
	Part 1	
23	Possession, completion	**Principal applicable clause**
26.2.5	Loss and expense	Architect's instructions regarding postponement as a cause of loss and expense

Clause No	Clause Title	Signpost
	Part 1	
17	Practical Compl./Defects	**Principal applicable clause**
1.3	Definitions etc.	Certificate of Completion of Making Good Defects
1.3	Definitions etc.	Defects Liability Period
1.3	Definitions etc.	Practical Completion
1.5	Definitions etc.	Contractor is responsible for the work irrespective of whether CPC issued
5.4.2	Contract Documents	Issue of information to the Contractor should have regard to likely date of PC
5.9	Contract Documents	Contractor to provide Employer with drawings showing PSW before the date of PC
18.1.1	Partial possession	Date of Practical Completion of Relevant Part
18.1.2	Partial possession	Architect to certify when defects in relevant part are made good
19.1.2	Assignment/Sub-Contracts	Right to bring proceedings may be assigned by Employer after issue of CPC
20.3.1	Injury and indemnity	'Property real or personal' excludes Works and Site Materials before issue of CPC
21.2.1.5	Insurance: persons etc	Insurance excludes damage to Works and Site Materials before issue of CPC
22.3.1	Insurance of the Works	Recognitions of NSCs within Joints Names Policy up to issue of CPC
22A.1	All risks - Contractor	Joint Names Policy to date of CPC
22B.1	All risks - Employer	Joint Names Policy to date of CPC

PRACTICAL COMPLETION

Clause No	Clause Title	Signpost
22C.1	Insurance - existing structures	Joint Names Policy to date of CPC – existing
22C.2	Insurance - existing structures	Joint Names Policy to date of CPC – Works in or extensions to existing
22D.1	Insurance: Loss of LADs	Insurance to date of CPC
23.3.1	Possession, completion	Contractor retains possession of site and Works to date of CPC
23.3.2	Possession, completion	With consent, Employer may occupy site or works or parts before date of CPC
24.2.1	Damages	For period between Completion Date and Practical Completion
25.3.3	Extension of time	Confirmation of existing or fixing new Completion Date
27.2.1	Determination by Employer	Identifying defaults made before date of Practical Completion
27.6.1	Determination by Employer	Employer may employ others to make good defects under cl.17
27.6.4.2	Determination by Employer	Employer's statement or Architect's certificate in relation to cl.17 defects
28.2.2	Determination by Contractor	Identifying events occurring before date of Practical Completion
28A.1.1	Determination by Contractor	Identifying events occurring before date of Practical Completion
30.1.3	Certificates/payments	Interim Certificates up to date of Practical Completion
30.2.3.1	Certificates/payments	Deductions not subject to Retention
30.4.1.2	Certificates/payments	Retention prior to Practical Completion
30.4.1.3	Certificates/payments	Retention after Practical Completion
30.6.1.1	Certificates/payments	Timing for sending documents for adjustment of Contract Sum

Clause No	Clause Title	Signpost
30.6.2.4	Certificates/payments	Deductions from Contract Sum
	Part 2	
35.16	Practical Completion	**Principal applicable clause**
35.19.1	Early Final Payment	Responsibility for loss up to Practical Completion of Works
35.24.5	Re-nomination	Circumstances where Architect exercises powers under cl.17
	Appendices	
17.2	Appendix	Defects Liability Period
19.1.2	Appendix	Assignment of benefits
30.1.3	Appendix	Dates of issue of Interim Certificates
2	Retention Bond	Reduction of maximum aggregate sum

PRICE STATEMENT

Clause No	Clause Title	Signpost
	Part 1	
13.4.1.2.A1-A7	Contractor's Price Statement	**Principal applicable clause**
1.3	Definitions etc.	Price Statement
1.3	Definitions etc.	Valuation
30.2.1.1	Certificates/payments	Accepted sums or part sums calculated as due in Interim Certificates
30.2.1.1	Certificates/payments	Amended accepted sums or part sums calculated as due in Interim Certificates
30.6.2	Certificates/payments	Accepted sums or part sums calculated as adjustment to Contract Sum
30.6.2	Certificates/payments	Amended accepted sums or part sums calculated as adjustment to Contract Sum

Clause No	Clause Title	Signpost
	Part 1	
1.3	Definitions etc.	Activity Schedule
13.4.2	Variations/Prov. Sums	Arising from instructions on expenditure of Provisional Sum
30.6.2.1	Certificates/payments	To be deducted in adjustment of Contract Sum
	Part 2	
35.1	General	Reservation of right to nominate
36.1.1.1	Nominated Suppliers	For supply of materials and goods
36.1.1.2	Nominated Suppliers	Instruction in regard to expenditure of Provisional Sum
36.1.1.3	Nominated Suppliers	Instruction in regard to expenditure of Provisional Sum
36.1.1.4	Nominated Suppliers	Arising out of a Variation
36.1.2	Nominated Suppliers	Application of the term 'Nominated Supplier'
36.2	Nominated Suppliers	Instructions to be issued nominating a supplier

PROVISIONAL SUMS

Clause No	Clause Title	Signpost
		Part 1
13	Variations/Prov. Sums	**Principal applicable clause**
1.3	Definitions etc.	Activity Schedule
1.3	Definitions etc.	Provisional Sum
2.2.2.2	Contractor's obligations	Errors in information shall be corrected as a Variation
5.4.2	Contract Documents	Architect to issue instructions with regard to the expenditure of Prov. Sums
5.9	Contract Documents	Contractor to provide as-built drawings for PSW the subject of Prov. Sums
6.2.3	Statutory obligations	Contractor to pay statutory charges that are stated by way of Prov. Sums
25.3.1.4	Extension of time	Regard given to expenditure of Prov. Sums in fixing a new Completion Date
25.3.2	Extension of time	Regard given to expenditure of Prov. Sums in fixing an earlier Completion Date
25.3.3.2	Extension of time	Regard given to expenditure of Prov. Sums in fixing an earlier Completion Date
25.4.5.1	Extension of time	Instructions on expenditure is a Relevant Event
26.2.7	Loss and expense	Instructions on expenditure may be grounds for claim
26.3	Loss and expense	Architect to notify Contractor of any extension of time related to cl.13.3
26.4.2	Loss and expense	Architect to notify Contractor of any extension of time for NSCs related to cl.13.3
30.2.1.5	Certificates/payments	Calculation of profit on nominated work
30.6.2.2	Certificates/payments	To be deducted in adjustment of Contract Sum

Clause No	Clause Title	Signpost
30.6.2.9	Certificates/payments	Calculation of profit on nominated work
30.6.2.12	Certificates/payments	Adjustment of Contract Sum
	Part 2	
35.1.2	General	Expenditure where rights of approval and selection reserved
35.2.1	General	Nomination for work the subject of an instruction issued under cl.13.3
36.1.1.2	Nominated Suppliers	Instructions on expenditure as Prime Cost Sum
36.1.1.3	Nominated Suppliers	Instructions on expenditure as Prime Cost Sum
36.4.3	Nominated Suppliers	Supply contract to allow for delivery programme to be varied by an instruction
	Part 5	
42.1.4	Performance Specified Work	Definition of Performance Specified Work
42.3	Performance Specified Work	Contractor's Statement to include for any instruction on the expenditure of a Prov. Sum
42.4	Performance Specified Work	Timing for provision of Contractor's Statement
42.7	Performance Specified Work	Definition of Provisional Sum for Performance Specified Work
42.8	Performance Specified Work	Only Prov. Sums stated in Contract bills as being for PSW can be the subject of PSW

QUANTITY SURVEYOR

Clause No	Clause Title	Signpost
Art. 4	Articles	Appointment of Quantity Surveyor
Part 1		
1.3	Definitions etc.	Quantity Surveyor
1.3	Definitions etc.	Valuation
5.1	Contract Documents	Custody of Contract Drawings and Bills
5.7	Contract Documents	Rates and prices in Contract Bills are not to be divulged
13.4.1.2.A1 (A)	Contractor's Price Statement	Contractor to submit Price Statement to Quantity Surveyor
13.4.1.2.A2	Contractor's Price Statement	Notifies Contractor within 21 days on whether or not Price Statement accepted
13.4.1.2.A4.1	Contractor's Price Statement	Notifies Contractor of reasons why Price Statement not accepted
13.4.1.2.A6	Contractor's Price Statement	Supplies an amended Price Statement that is acceptable
13.4.1.2.A7.1	Contractor's Price Statement	Notifies Contractor within 21 days on whether or not cls. A1.1 requirements accepted
13.4.1.2.A7.2	Contractor's Price Statement	Cls. 25 + 26 apply if Contractor not notified within 21 days
13.4.1.2 (B)	Variations/Prov. Sums	Valuation to be in accordance with cls.13.5.1 - .7
13.6	Variations/Prov. Sums	Opportunity for Contractor to witness measurements
13A.1.2	Variation instruction	Contractor to submit 13A Quotation to Quantity Surveyor
13A.1.2	Variation instruction	Timing for submission of 13A Quotation

Clause No	Clause Title	Signpost
26.1	Loss and expense	Ascertainment of Contractor's loss and expense if instructed
26.1.3	Loss and expense	Submission of details of loss and expense
26.4.1	Loss and expense	Ascertainment of NSC's loss and expense if instructed
30.1.2.1	Certificates/payments	Interim Valuations when Architect considers them necessary
30.1.2.2	Certificates/payments	Contractor may submit an application to the Quantity Surveyor
30.1.2.2	Certificate /payments	Quantity Surveyor to identify any disagreement with the application
30.6.1.1	Certificate /payments	May be sent documents for adjustment of Contract Sum
30.6.1.2.1	Certificates/payments	Ascertainment of loss and expense; timing
30.6.1.2.2	Certificates/payments	Preparation of a statement of all adjustments to the Contract Sum; timing

Part 2

35.17.2	Early Final Payment	Documents for final adjustment of Sub-Contract Sum

Part 3

38.4.3	Sub-let/Domestics	Agreement of net amount payable or allowable
38.4.5	Sub-let/Domestics	Contractor to provide evidence reasonably required
39.5.3	Sub-let/Domestics	Agreement of net amount payable or allowable
39.5.5	Sub-let/Domestics	Contractor to provide evidence reasonably required
40.5	Price adjustment formulae	Agreement of any alteration to methods of ascertainment

13A QUOTATION

Clause No	Clause Title	Signpost
25.3.5	Extension of time	Contractor to notify NSCs of any change of Completion Date
25.4.5.1	Extension of time	Acceptance of 13A Quotation is not a Relevant Event
26.2.7	Loss and expense	Acceptance of 13A Quotation is not a matter for loss and expense claims
30.6.2	Certificates/payments	Adjustment of Contract Sum

REASONABLE

Clause No	Clause Title	Signpost
Art. 3	Articles	Nomination of a replacement Architect within a reasonable time
Art. 4	Articles	Nomination of a replacement Quantity Surveyor within a reasonable time
	Part 1	
2.1	Contractor's obligations	Satisfaction of Architect as to quality and standards
4.1.1.1	Architect's Instructions	Objection to Variation made under cl.13.1.2
5.1	Contract Documents	Times for inspection of Documents
5.4.1	Contract Documents	Consent to vary times given in Information Release Schedule
5.4.2	Contract Documents	Drawings necessary to amplify Contract Drawings
5.4.2	Contract Documents	Timing of issue of further drawings etc. having regard to Completion Date
5.4.2	Contract Documents	Grounds for believing that Architect is not aware of the times drawings were needed
5.4.2	Contract Documents	Advise the Architect sufficiently in advance of a requirement for further drawings etc
5.5	Contract Documents	Availability of documentation on site
6.1.4.1	Statutory obligations	Work and materials necessary in an emergency
6A.3	CDM Regulations	Contractor to comply with requirements of Principal Contractor
6A.4	CDM Regulations	Time Planning Supervisor requires to provide information for the health and safety file
8.1.1	Work, mats. and goods	Materials and goods to satisfaction of the Architect

Clause No	Clause Title	Signpost
8.1.2	Work, mats. and goods	Workmanship to satisfaction of the Architect
8.1.4	Work, mats. and goods	Consent to substitution of materials or goods
8.2.2	Work, mats. and goods	Expression of dissatisfaction of Architect with materials, goods, workmanship
8.2.2	Work, mats. and goods	Time for Architect to express dissatisfaction with materials, goods, workmanship
8.4.3	Work, mats. and goods	Instructions for Variation as a consequence of instructions under cls.8.4.1 or .2
8.4.4	Work, mats. and goods	Instructions for opening up for inspection
8.4.4	Work, mats. and goods	Satisfaction of the Architect as to extent of non-compliance
8.4.4	Work, mats. and goods	No addition to the Contract Sum
8.5	Work, mats. and goods	Instructions necessary as a consequence of non-compliance with cl. 8.1.3
8.6	Work, mats. and goods	Instructions requiring the exclusion from the site of any person employed
11	Access to the Works	Times for access
11	Access to the Works	Securing right of access for Architect to Sub-Contractor workshops
12	Clerk of Works	Facilities for performance of duties
13.2.2	Variations/Prov. Sums	Right of objection to instruction under cl.13.2.1
13.5.1.4	Variations/Prov. Sums	Rates used to value work for which Approximate Quantities in Contract Bills
13.5.1.5	Variations/Prov. Sums	Rates used to value work for which Approximate Quantities in Contract Bills
13.5.7	Variations/Prov. Sums	Circumstances for fair valuation being made

REASONABLE

Clause No	Clause Title	Signpost
13A.1.1	Variation instruction	Sufficiency of information provided to prepare 13A Quotation
13A.2.4	Variation instruction	Cost of preparing 13A Quotation
13A.2.6	Variation instruction	Sufficient supporting information with each part of 13A Quotation
13A.5	Variation instruction	Amount added to Contract Sum for preparation of 13A Quotation
13A.5	Variation instruction	Addition to Contract Sum only if 13A Quotation prepared on fair basis
13A.5	Variation instruction	Non-acceptance of 13A Quotation is not evidence of unfair basis of preparation
13A.8	Variation instruction	Evaluation of Variations to work covered by 13A Quotation
16.1	Mats. and goods unfixed	Consent to removal of goods or materials
17.2	Practical Compl./Defects	Time for making good defects
17.3	Practical Compl./Defects	Time for making good defects
18.1	Partial Possession	Consent of Contractor to Employer taking partial possession
19.2.2	Assignment/Sub-Contracts	Consent to sub-letting
19.3.2.1	Assignment/Sub-Contracts	Additions to list for named firms
19.3.2.2	Assignment/Sub-Contracts	Additions to list for named firms
19.4.2.1	Assignment/Sub-Contracts	Consent to removal of materials from site by a Sub-Contractor
21.1.2	Insurance: persons etc	Requirement for production of documentary evidence of insurance
21.1.2	Insurance: persons etc	Requirement for production of insurance policies and receipts

Clause No	Clause Title	Signpost
21.2.1.3	Insurance: persons etc.	Damage to property other than the Works
22.2	Insurance of the Works	All Risks insurance to cover removal and disposal
22A.3.1	All Risks – by Contractor	Requirement for production of documentary evidence of insurance
22A.3.1	All Risks – by Contractor	Requirement for production of insurance policies and receipts
22B.2	All Risks – by Employer	Requirement for production of documentary evidence of insurance
22C.3	Insurance - existing structures	Requirement for production of documentary evidence of insurance
22D.1	Insurance: loss of LADs	Architect to obtain from Employer information for Contractor to obtain quotation
22D.1	Insurance: loss of LADs	Timing of instruction whether or not quotation is to be accepted
22FC.3.2	Joint Fire Code	Failure regularly and diligently to proceed with Remedial Measures
23.3.2	Possession, completion	Consent of Contractor for Employer to occupy site or Works before issue of CPC.
25.2.1.1	Extension of time	Timing for notice of circumstances causing delay
25.2.3	Extension of time	Further notices necessary or required by the Architect
25.3.1	Extension of time	Fixing of a later Completion Date
25.3.1	Extension of time	Timing for fixing of a new Completion Date
25.3.2	Extension of time	Fixing of an earlier Completion Date
25.3.3.1	Extension of time	Timing for subsequent fixing of a later Completion Date
25.3.3.2	Extension of time	Timing for subsequent fixing of an earlier Completion Date
25.3.4.2	Extension of time	Contractor's obligation to proceed with the Works

REASONABLE

Clause No	Clause Title	Signpost
25.4.10.1	Extension of time	Inability to foresee difficulties in obtaining labour
25.4.10.2	Extension of time	Inability to foresee difficulties in obtaining goods and materials
25.4.14	Extension of time	Work, the subject of insufficiently accurate Approximate Quantities
26.1.1	Loss and expense	Timing for application to be made
26.1.2	Loss and expense	Information to allow Architect to form an opinion
26.1.3	Loss and expense	Details necessary for ascertainment
26.2.8	Loss and expense	Work, the subject of insufficiently accurate Approximate Quantities
27.2.1.1	Determination by Employer	Suspension of Works
27.2.3	Determination by Employer	Timing of repetition of specified default
27.2.4	Determination by Employer	Notice of determination
27.5.4	Determination by Employer	Employer's measures to ensure Site Materials, site and Works are protected
27.5.4	Determination by Employer	Deduction of cost of taking measures to protect
27.6.3	Determination by Employer	Timing for Contractor to remove his property
27.6.4.1	Determination by Employer	Discharge of amounts already properly due to the Contractor
27.6.4.2	Determination by Employer	Timing for issue of account in respect of matters detailed in cl.27.5.6
27.7.1	Determination by Employer	Timing for issue of statement of account when Employer abandons the Works.
28.2.4	Determination by Contractor	Timing of repetition of specified default
28.2.5	Determination by Contractor	Notice of determination

Clause No	Clause Title	Signpost
28.4.1	Determination by Contractor	Timing for Contractor to remove his property
28.4.3	Determination by Contractor	Timing for issue of account in respect of matters detailed in cls.28.4.3.1 - .5
28.4.3.3	Determination by Contractor	Removal costs which may be added to amounts payable
28A.1.3	Determination by Contractor	Notice of determination
28A.3	Determination by Contractor	Timing for Contractor to remove his property
28A.5.3	Determination by Contractor	Removal costs which may be added to amounts payable
28A.7	Determination by Contractor	Amounts attributable to NSCs
29.2	Works by Employer	Consent to work not forming part of the Contract
30.2.1.2	Certificates/payments	Conditions for value of materials being included
30.2.1.5	Certificates/payments	Rates at which profit calculated on nominated work
30.3.1	Certificates/payments	Proof that property of listed items lies in the Contractor
30.3.2	Certificates/payments	Proof that property of listed items lies in the Contractor
30.3.5	Certificates/payments	Proof that listed items are insured
30.6.2.9	Certificates/payments	Rates at which profit calculated on nominated work
30.9.1.1	Certificates/payments	Final Certificate as conclusive evidence of quality etc
31.5.3	CIS	Grounds for the Employer to believe that a cl.31.5.1.1 statement is incorrect

REASONABLE

Clause No	Clause Title	Signpost
35.24.10	Re-nomination	Timing for further nomination
36.4.1	Nominated Suppliers	Satisfaction of the Architect as to quality and standards
36.4.2	Nominated Suppliers	Nominated Supplier bearing expenses arising from defects
36.4.2.1	Nominated Suppliers	Examination for defects before fixing
36.4.3	Nominated Suppliers	Failure to provide information as grounds for varying agreed delivery programme
36.4.3	Nominated Suppliers	Deliveries in accordance with directions of the Contractor
	Part 3	
38.4.2	Clause 38 provisions	Timing of notice under cl.38.4.1
38.4.5	Clause 38 provisions	Timing for Contractor to provide evidence and computations
38.4.5	Clause 38 provisions	Evidence and computations required by the Architect or Quantity Surveyor
38.4.5	Clause 38 provisions	Certificate for validity of the evidence
39.5.2	Clause 39 provisions	Timing of notice under cl.39.5.1
39.5.5	Clause 39 provisions	Timing for Contractor to provide evidence and computations
39.5.5	Clause 39 provisions	Evidence and computations required by the Architect or Quantity Surveyor
39.5.5	Clause 39 provisions	Certificate for validity of the evidence
40.5.1	Price adjustment formulae	Conditions under which alteration of methods and procedures may be agreed
40.6.1	Price adjustment formulae	Basis upon which adjustment is to be made in the absence of Monthly Bulletins

REASONABLE

Clause No	Clause Title	Signpost
	Part 4	
41A.5.8	Adjudication	Dispute as to whether an instruction is reasonable
41A.5.8.2	Adjudication	Power to appoint independent expert on whether an instruction is reasonable
41A.6.1	Adjudication	Statement on apportionment of Adjudicator's expenses
41A.6.1	Adjudication	Apportionment in the absence of a statement
41A.6.2	Adjudication	Joint and several liability for Adjudicator's expenses
	Part 5	
42.3	Performance Specified Work	Timing of Contractor's Statement in relation to further information from Architect
42.4	Performance Specified Work	Timing of issue of Contractor's Statement
42.14	Performance Specified Work	Timing of instructions for integration of PSW with the Works
42.15	Performance Specified Work	Consent of Contractor to instruction for removal of injurious affection
42.17.1	Performance Specified Work	Degree of skill and care in provision of PSW
	Appendices	
1	Code of Practice: Clause 8.4.4	Purpose of Code
2.12	Code of Practice: Clause 8.4.4	Likelihood of similar non-compliance
8	Advance Payment Bond	Consent of Surety to transfer or assignment

Clause No	Clause Title	Signpost
10	Off-site Materials etc. Bond	Consent of Surety to transfer or assignment
VAT 1.2.1	VAT Agreement	Objection by the Employer to the calculation of values
VAT 1.2.2	VAT Agreement	Procedure for objection to provisional assessment
VAT 7.2	VAT Agreement	Circumstances in which Employer is not obliged to make further payments
5	Retention Bond	Consent of Surety to transfer or assignment

RELEVANT EVENT

Clause No	Clause Title	Signpost
	Part 1	
25.4	Extension of time	**Principal applicable clause**
1.3	Definitions etc.	Relevant Event
25.2.1.1	Extension of time	Contractor to give written notice identifying
25.2.2.2	Extension of time	Particulars of effect and estimated delay for every Relevant Event
25.3.1.1	Extension of time	Circumstances under which an extension given
25.3.1.3	Extension of time	Statement of Relevant Event taken into account
25.3.3.1	Extension of time	Timing and circumstances for fixing a later Completion Date
26.3	Loss and expense	Statement of extension as necessary for ascertainment
26.4.2	Loss and expense	Statement of extension under relevant clauses of NSC/C
34.3.2	Antiquities	Statement of extension given in respect of cl.25.4.5.1
	Part 5	
42.16	Performance Specified Work	Restrictions on granting of extensions of time

Clause No	Clause Title	Signpost
	Part 1	
30.4; 30.4A; 30.5	Certifcates/payments	**Principal applicable clauses**
1.3	Definitions etc	Retention
1.3	Definitions etc	Retention Percentage
18.1.1	Partial possession	Practical Completion and Defects Liability Period start of the Relevant Part
22A.5.2.2	All Risks – Contractor	Provisions requiring release of Retention do not apply on determination
22B.4.2.2	All Risks – Employer	Provisions requiring release of Retention do not apply on determination
22C.1A.3	Insurance - existing structures	Provisions requiring release of Retention do not apply on determination
22C.5.2.2	Insurance - existing structures	Provisions requiring release of Retention do not apply on determination
27.6.4.1	Determination by Employer	Provisions requiring release of Retention do not apply on determination
28.4	Determination by Contractor	Provisions requiring release of Retention do not apply on determination
28.4.2	Determination by Contractor	Timing for payment by Employer of Retention deducted
28.4.3	Determination by Contractor	Payments properly due without deduction of Retention
28A.2	Determination by Contractor	Provisions requiring release of Retention do not apply on determination
28A.4	Determination by Contractor	Timing for payment of Retention
28A.5	Determination by Contractor	Payments properly due without deduction of Retention

RETENTION

Clause No	Clause Title	Signpost
30.1.1.2	Certificates/payments	Entitlement to make deductions whether or not Retention held
30.2	Certificates/payments	Deducted from amounts of gross Valuation
30.2.1	Certificates/payments	List of items included subject to Retention
30.2.2	Certificates/payments	List of items included not subject to Retention
30.2.3	Certificates/payments	List of items deducted not subject to Retention
	Part 2	
35.13.5.3.2	Payment	Reduction in cl.35.13.5.2 not to exceed Contractor's Retention
	Appendices	
30.4.1.1	Appendix	Retention Percentage
30.4A	Appendix	Retention Bond
30.4A.2	Appendix	Retention Bond – amount and date of expiry
VAT 1.1	VAT Agreement	Provisional written assessment of values less Retention

Clause No	Clause Title	Signpost
	Part 1	
21.2.4	Insurance: persons etc.	Employer may insure against any risk if Contractor defaults
22A.2	All Risks – Contractor	Employer may insure against any risk if Contractor defaults
22A.4.1	All Risks – Contractor	Procedure if loss or damage occasioned by any of the risks
22B.2	All Risks – Employer	Contractor may insure against any risk if Employer defaults
22B.3.1	All Risks – Employer	Procedure if loss or damage occasioned by any of the risks
22C.3	Insurance - existing structures	Contractor may insure against any risk if Employer defaults under cl.22C.1
22C.3	Insurance - existing structures	Contractor may insure against any risk if Employer defaults under cl.22C.2
22C.4	Insurance - existing structures	Procedure if loss or damage occasioned by any of the risks
22D.4	Insurance for loss of LADs	Employer may insure against any risk if Contractor defaults

ROYALTIES AND PATENT RIGHTS

Clause No	Clause Title	Signpost
	Part 1	
9	Royalties etc.	**Principal applicable clause**
30.2.2.1	Certificates/payments	Amounts included not subject to Retention
30.6.2.10	Certificates/payments	Amounts defined in cl.9.2 added to Contract Sum

Clause No	Clause Title	Signpost
	Part 1	
1.3	Definitions etc.	Specified Perils
22.3.1	Insurance of the Works	Joint Names Policy under cls.22A/B/C to include specific items related to NSCs
22C.1	Insurance - existing structures	Employer to take out Joint Names Policy in respect of damage to existing buildings
22D.1	Insurance: Loss of LADs	When requested, Contractor to obtain insurance against Specified Perils
25.4.3	Extension of time	Loss or damage caused – Relevant Event
28A.1.1.2	Determination by either	Loss or damage caused – ground for determination
28A.1.2	Determination by either	Loss or damage caused - not a ground for determination if caused by negligence etc.
28A.6	Determination by either	Accounts to include loss or damage where cl.28A.1.1.2 is fault of Employer
30.3.5	Certificates/payments	Insurance of off-site materials during period of from transfer of property to delivery

STATUTORY REQUIREMENTS

Clause No	Clause Title	Signpost
	Part 1	
6	Statutory obligations	**Principal applicable clause**
1.3	Definitions etc.	Statutory Requirements
25.4.11	Extension of time	Work in pursuance of Statutory Requirements as Relevant Event
25.4.15	Extension of time	Change in Statutory Requirements that alters PSW as a Relevant Event
30.2.2.1	Certificates/payments	Amounts not subject to Retention
30.6.2.10	Certificates/payments	Amounts as defined in cl.6.2 added to Contract Sum
	Appendices	
2.4	Code of Practice: Clause 8.4.4	The effect of non-compliance on the need to comply with Statutory Requirements

Clause No	Clause Title	Signpost
	Part 1	
13.4.2	Variations/Prov. Sums	Valuation of work tendered by Contractor
30.6.2.6	Certificates/payments	Contract Sum adjustment to include final sums for NSCs
30.6.2.7	Certificates/payments	Contract Sum adjustment to include adjusted sum under cl.35.2
	Part 2	
35.2	General	**Principal applicable clause**
	Appendices	
35.2	Appendix	Work for which Contractor desires to tender

VALUATION

Clause No	Clause Title	Signpost
	Part 1	
13.4; 13.5	Variations/Prov. Sums	**Principal applicable clauses**
30.2	Certificates/payments	
1.3	Definitions etc.	Valuation
1.3	Definitions etc.	Price Statement
1.3	Definitions etc.	Activity schedule
1.3	Definitions etc.	Retention
8.4.3	Work, materials and goods	Where goods or materials are not in accordance with Contract
8.5	Work, materials and goods	Failure to comply with cl.8.1.3
13.2.3	Variations/Prov. Sums	Valuation of Variations
13.6	Variations/Prov. Sums	Opportunity for Contractor to witness measurements
13.7	Variations/Prov. Sums	Adjustment of Contract Sum gives effect to Valuation
16.1	Mats. And goods unfixed	Unfixed materials included in an Interim Certificate
19.4.2.2	Assignment/Sub-Contracts	Provisions to be included in a Sub-Contract for property in materials
19.4.2.3	Assignment/Sub-Contracts	Provisions to be included in a Sub-Contract for property in materials
25.3.2	Extension of time	Fixing of an earlier Completion Date
30.1.2.1	Certificates/payments	Valuations by the Quantity Surveyor whenever the Architect considers them necessary

Clause No	Clause Title	Signpost
30.1.2.2	Certificates/payments	Contractor may submit to the Quantity Surveyor an application for gross valuation
30.4.1	Certificates/payments	Calculation of Retention
30.4.1.2	Certificates/payments	Amounts from which Retention deducted
30.4A.1	Certificates/payments	Retention Bond
30.4A.3	Certificates/payments	Application of normal rules where Contractor fails to provide Retention Bond
30.4A.4	Certificates/payments	Procedure where Retention Bond cover not adequate
30.6.2	Certificates/payments	Adjustment of Contract Sum
30.6.2.3	Certificates/payments	Valuation of items omitted in adjustment of Contract Sum
30.6.2.11	Certificates/payments	Value of Variations added in adjustment of Contract Sum
30.6.2.12	Certificates/payments	Work for Provisional Sums in adjustment of Contract Sum
30.6.17	Certificates/payments	Loss and expense accepted as part of a Price Statement
30.7	Certificates/payments	Interim Certificate including a gross valuation of Sub-Contractors
	Part 2	
35.17	Early Final Payment	Interim Certificate including a gross valuation of Sub-Contractors

VALUATION

Clause No	Clause Title	Signpost
	Part 1	
13.1; 13.2	Variations/ Prov. Sums	**Principal applicable clauses**
1.3	Definitions etc.	Variation
2.2.2.2	Contractor's obligations	Errors in Bills or deviation from Method of Measurement
2.3.3	Contractor's obligations	Divergence between instruction and Contract Documents
4.1.1.1	Architect's Instructions	Reasonable objection to an instruction
6.1.2	Statutory obligations	Contractor action at divergence between instruction and Statutory Requirements
6.1.3	Statutory obligations	Architect action on receiving notification from Contractor
6.1.3	Statutory obligations	Instructions requiring works to be varied
6.1.4.3	Statutory obligations	Treatment of emergency work as a Variation
6.1.5	Statutory obligations	Non -liability of Contractor if instructed work does not conform
6.1.7	Statutory obligations	Statutory Requirements necessitating change in Performance Specified Work
8.4.2	Work, materials and goods	Work not in accordance with the Contract accepted but not treated as a Variation
8.4.3	Work, materials and goods	Work not in accordance with the Contract treated as a Variation
8.5	Work, materials and goods	Work not carried out in a proper and workmanlike manner as a Variation
13.4.1.1	Variations/Prov. Sums	Valuation of Variations
13.4.1.3	Variations/Prov. Sums	Variations to Sub-Contract Works valued in accordance with NSC/C

VARIATION

Clause No	Clause Title	Signpost
13.5.5	Variations/Prov. Sums	Change of conditions under which other work executed
13.5.6	Variations/Prov. Sums	Change of conditions under which Performance Specified Work executed
13.5.7	Variations/Prov. Sums	Circumstances for fair valuation being made
22A.5.3	All risks - Contractor	Restoration of work damaged
22B.3.5	All risks - Employer	Restoration of work damaged
22B.4.3	All risks - Employer	Restoration of work damaged
22C.4.4.2	Insurance - existing structures	Restoration of work damaged
22C.5.3	Insurance - existing structures	Restoration of work damaged
25.3.1.4	Extension of time	Statement of the effect of omissions on Completion Date
25.3.2	Extension of time	Account to be taken of omissions in fixing Completion Date
25.3.3.2	Extension of time	Account to be taken of omissions in fixing Completion Date
25.4.5.1	Extension of time	Instruction requiring a Variation is a Relevant Event
26.2.7	Loss and expense	Instruction requiring a Variation may be a ground for claim
26.3	Loss and expense	Architect to state extension of time in writing for ascertaining loss and expense
28.2.2.2	Determination by Contractor	Instruction causing suspension of the Works for period longer than stated in Appendix
28A.1.1.4	Determination by either	Architect instruction as a result of negligence by local authority etc.
30.2.1.1	Certificates/payments	Work treated as a Variation under cls. 22B + 22C not subject to Retention
30.2.2.2	Certificates/payments	Work treated as a Variation under cls. 22B + 22C not subject to Retention

Clause No	Clause Title	Signpost
30.6.2	Certificates/payments	Contract Sum adjusted by amount of agreed cl. 13.4.1.1 valuations
30.6.2.3	Certificates/payments	Valuation of omissions in adjustment of Contract Sum
30.6.2.11	Certificates/payments	Valuation of Variations in adjustment of Contract Sum
Part 2		
35.1.3	General	Instructions which reserve rights to nomination
35.5.2	Procedure	Omission of work the subject of a nomination instruction to which objection was made
35.9.2	Procedure	Omission of work the subject of a nomination instruction to which objection was made
36.1.1.4	Nominated Suppliers	Causing expenditure in favour of a sole supplier
36.4.3	Nominated Suppliers	Conditions upon which supply contracts must be let to allow for Variations
Part 5		
42.10	Performance Specified Work	Correction of information in the contract Bills
42.11	Performance Specified Work	Architect may issue instructions for Variations to PSW
42.12	Performance Specified Work	Without agreement Architect may not instruct PSW extra to that identified in Appendix

VAT

Clause No	Clause Title	Signpost
	Part 1	
8	Materials, goods etc.	**Principal applicable clause**
2.1	Contractor's obligations	Quality and standards specified in Contracts Documents
2.1	Contractor's obligations	Where a matter for his opinion, quality to be to Architect's satisfaction
17.2	Practical Compl./Defects	Defects appearing in Defects Period are to be specified
17.3	Practical Compl./Defects	Defects at any time may be the subject of an instruction
22.2.2	Insurance of the Works	Defective workmanship excluded from All-Risks Insurance
25.4.5.2	Extension of time	Checking of workmanship found to be in accordance with Contract
26.2.2	Loss and expense	Checking of workmanship found to be in accordance with Contract
30.2.2.1	Certificates/payments	Test and inspection costs added to the amounts payable
30.2.3.1	Certificates/payments	Deductions from sums payable for work not in accordance with Contract
30.6.2.4	Certificates/payments	Deductions from Contract Sum for work not in accordance with Contract
30.6.2.10	Certificates/payments	Test and inspection costs added to the Contract Sum
30.9.1.1	Certificates/payments	Final Certificate as conclusive evidence of quality

WORKMANSHIP

Clause No	Clause Title	Signpost
	Part 2	
35.21	Clause 2.1 of NSC/W	Nothing in cl. 35.21 affects obligations of Contractor
35.24.5	Re-nomination	Where NSC cannot be required or does not agree to re-execute certain work
36.4.1	Nominated Suppliers	Quality to be to satisfaction of the Architect
36.4.2.2	Nominated Suppliers	Defects due to defective workmanship
	Part 4	
41A.5.8	Adjudication	Disputes arising under cl.8.4.4
41A.5.8.2	Adjudication	Appointment of an expert in disputes arising under cl.8.4.4
	Part 5	
42.17.1.1	Performance Specified Work	Cl.42.17 does not affect obligations of Contractor regarding workmanship
	Appendices	
2.2	Clause 8.4.4 Code of Practice	Failure of workmanship as a cause of non-compliance

Clause No	Clause Title	Signpost
	Part 1	
1.6	Definitions etc.	Notifying Contractor of replacement Planning Supervisor or Principal Contractor
1.9	Definitions etc.	Notifying Contractor of appointment of Employer's Representative
2.3	Contractor's obligations	Discrepancies between Contract Documents
2.4.1	Contractor's obligations	Notifying Architect of discrepancy between Contractor's Statement and any instruction
2.4.2	Contractor's obligations	Notifying Architect of any discrepancy within Contractor's Statement
4.1.1.1	Architect's Instructions	Reasonable objection to an instruction
4.1.2	Architect's Instructions	Failure to comply with notice requiring compliance
4.2	Architect's Instructions	Request for statement of authority to issue instructions
4.3.1	Architect's Instructions	All instructions
4.3.2	Architect's Instructions	Instructions not in writing may subsequently be confirmed in writing
4.3.2	Architect's Instructions	Dissent by Architect to Contractor's confirmation
4.3.2.1	Architect's Instructions	Confirmation by Architect of instructions not in writing
4.3.2.2	Architect's Instructions	Confirmation at any time prior to Final Certificate
6.1.2	Statutory obligations	Divergence of Contract Documents from Statutory Requirements
6.1.6	Statutory obligations	Architect or Contractor to notify the other of divergence from Contractor's Statement
6.1.6	Statutory obligations	Contractor to notify Architect of proposed amendment

WRITING

Clause No	Clause Title	Signpost
6A.4	CDM Regulations	Information for health and safety file within time required by Planning Supervisor
8.1.4	Work, materials and goods	Architect's consent to substitution of materials for PSW
8.4.2	Work, materials and goods	Confirmation of an allowance for non-compliant work etc. to remain
12	Clerk of Works	Confirmation of Clerk of Works directions
13.2.3	Variations/Prov. Sums	Contractor's disagreement with application of cl.13A to an instruction
13.2.4	Variations/Prov. Sums	Sanctioning of Variations made by Contractor
13.4.1.1	Variations/Prov. Sums	Method of valuation of sanctioned Variations
13.4.1.2.A2	Contractor's Price Statement	Notification by Quantity Surveyor of decision on Price Statement
13A.3.1	Variation instruction	Notification by Employer of acceptance of 13A Quotation
13A.3.2	Variation instruction	Confirmation by Architect of the Employer's acceptance
13A.7	Variation instruction	Increase or reduction in number of days for providing a 13A Quotation
16.1	Mats. and goods unfixed	Removal of unfixed materials on site
18.1	Partial possession	Architect issues statement identifying part and dates
19.1.1	Assignment/Sub-Contracts	Consent to assignment
19.2.2	Assignment/Sub-Contracts	Consent to sub-letting
19.4.2.1	Assignment/Sub-Contracts	Sub–contracts to provide that no materials on site be removed without written consent
22A.4.1	All Risks – Contractor	Contractor to notify Architect and Employer of loss or damage

Clause No	Clause Title	Signpost
22A.5.2	All Risks – Contractor	Employer to notify Contractor after receipt of the Insurers' Notification
22B.3.1	All Risks – Employer	Contractor to notify Architect and Employer of loss or damage
22B.4.2	All Risks – Employer	Employer to notify Contractor after receipt of the Insurers' Notification
22C.1A.1	Insurance - existing structures	One party to notify the other after receipt of the Insurers' Notification
22C.4	Insurance - existing structures	Contractor to notify Architect and Employer of loss or damage
22C.5.2	Insurance - existing structures	Employer to notify Contractor after receipt of the Insurers' Notification
23.3.2	Possession, completion	Consent of Contractor for Employer to use or occupy site or Works
24.2.1	Damages	Establishment of employer's right to liquidated damages
24.2.1.1	Damages	Requirement for Contractor to pay liquidated damages
24.2.1.1	Damages	Statement of rate of liquidated damages
24.2.3	Damages	Requirements of the Employer under cl.24.1 remain effective unless withdrawn
25.2.1.1	Extension of time	Notice of cause of delay
25.2.1.2	Extension of time	Copy of cl.25.2.1.1 notice to NSC where relevant
25.2.2	Extension of time	Notification of effect and estimated delay
25.2.2	Extension of time	Copy of particulars under 25.2.2 to NSC where relevant
25.2.3	Extension of time	Further notices to Architect to update particulars with copies to NSCs
25.3.1	Extension of time	Fixing revised Completion Dates
25.3.1.2	Extension of time	Notification of rejection of request for revised Completion Date

WRITING

Clause No	Clause Title	Signpost
25.3.3	Extension of time	Fixing revised Completion Dates
25.3.5	Extension of time	Notifications to NSCs of confirmation of Completion Date
26.1	Loss and expense	Applications for loss and expense
26.3	Loss and expense	Statement of extension given under cl.25
26.4.1	Loss and expense	Applications under cl.4.38.1 of NSC/C
26.4.2	Loss and expense	Statement of extension given under relevant clauses of NSC/C
27.1	Determination by Employer	Notices referred to in the Clause in writing
27.2.1.3	Determination by Employer	Refusal to comply with written notice as ground for determination
27.3.2	Determination by Employer	Contractor to notify Employer of any composition or arrangement with creditors etc.
27.6.3	Determination by Employer	Notice requiring removal of buildings, plant or material
27.7.1	Determination by Employer	Employer notifies Contractor is he does not intend to complete the Works
27.7.1	Determination by Employer	Employer to send Contractor a statement of account
27.7.2	Determination by Employer	Procedure if Employer does not give written notice under cl.27.7.1
28.1	Determination by Contractor	Notices referred to in the Clause in writing
28.3.2	Determination by Contractor	Contractor to notify Employer of any composition or arrangement with creditors etc.
28.1.3.5	Determination by Contractor	Non-receipt of instructions, drawings or details applied for
28A.1.1	Determination by either	Employer or Contractor may give the other notice of determination
28A.7	Determination by either	Proportion of payments that are attributable to NSCs

Clause No	Clause Title	Signpost
30.1.1.3	Certificates/payments	Notice from Employer specifying amount to be paid
30.1.1.4	Certificates/payments	Reason for exercising right of deduction from monies due
30.1.1.5	Certificates/payments	Payment of amount pursuant to cl.30.1.1.1 to be made if written notices not given
30.1.4	Certificates/payments	Contractor issues notice of suspension of the Works
30.6.2.3	Certificates/payments	Amounts deducted in adjustments to Contract Sum
30.8.2	Certificates/payments	Notice to Contractor specifying proposed payment and its basis
30.8.3	Certificates/payments	Notice to Contractor of proposed amounts to be withheld
30.8.4	Certificates/payments	Payment of amount in Final Certificate to be made if written notices not given
31.4.1	CIS	Dissatisfaction with the validity of the Authority provided by the Contractor
34.3.2	Antiquities	Statement of extension of time made
	Part 2	
35.5.1	Procedure	Reasonable objection to proposed NSC
35.8	Procedure	Contractor to inform Architect of alternatives following failure to comply with cl.35.7
35.9.2	Procedure	Architect response to notice issued under cl.35.8.2
35.13.6.1	Payment	Employer to notify Contractor of amounts to be credited to NSCs
35.14.1	Extension of Period	Consent to the granting of an extension by the Contractor
35.14.2	Extension of Period	Procedure upon receipt by Architect of request for consent

WRITING

Clause No	Clause Title	Signpost
35.15.1	Failure to complete	Within extended time granted with written consent
35.15.1	Failure to complete	To be certified in writing
35.16	Practical Completion	Procedure when Practical Completion by NSC achieved
35.24.2	Re-nomination	Determination of employment of NSC to be with written consent of Architect
35.24.7.1	Re-nomination	Conditions under which Architect may withhold consent
35.24.7.3	Re-nomination	Procedure where written consent to determination has been given
36.1.1.4	Nominated Suppliers	Nomination by written sanction
36.4.3	Nominated Suppliers	Supply contract to allow programme changes for delayed information issue.
36.5.1	Nominated Suppliers	Approval by Architect of restrictions, limitations etc.
36.5.2	Nominated Suppliers	Contractor not obliged to enter into contract before approval
Part 3		
38.4.1	Clause 38 provisions	Contractor to give Architect notice of occurrence of events
38.4.2	Clause 38 provisions	Written notice is condition precedent to any payment being made
38.4.8.2	Clause 38 provisions	Response by Architect to written notices under cl.25
38.4.8.2	Clause 38 provisions	Architect to have fixed new Completion Date in writing
39.5.1	Clause 39 provisions	Contractor to give Architect notice of occurrence of events
39.5.2	Clause 39 provisions	Written notice is condition precedent to any payment being made

Clause No	Clause Title	Signpost
39.5.8.2	Clause 39 provisions	Response by Architect to written notices under cl.25
39.5.8.2	Clause 39 provisions	Architect to have fixed new Completion Date in writing
40.7.2.2	Price adjustment formulae	Response by Architect to written notices under cl.25
40.7.2.2	Price adjustment formulae	Architect to have fixed new Completion Date in writing

Part 4

41A.5.2	Adjudication	Party not making the referral may send a statement of contentions
41A.5.3	Adjudication	Adjudicator decision
41A.5.5.3	Adjudication	Adjudicator may request further information beyond written statement
41A.5.8.2	Adjudication	Any independent expert appointed is to report in writing
41A.5.8.4	Adjudication	Any written advice from and independent expert to be supplied to the Parties
41A.7.1	Adjudication	Adjudicator's decision binding unless the Parties agree in writing otherwise
41B.1.1	Arbitration	Arbitration begins when one party serves a written notice on the other
41B.6	Arbitration	Rules for adoption of amended CIMAR rules

Part 5

42.5	Performance Specified Work	Architect in writing may require the Contractor to amend his Statement
42.15	Performance Specified Work	Notice to the Architect of injurious affection

WRITING

Clause No	Clause Title	Signpost
42.15	Performance Specified Work	Instruction does not have effect without the consent of the Contractor
	Appendices	
6	Advance Payment Bond	Advice by Employer to Surety of any reduction of the sum
7	Advance Payment Bond	All claims under the Bond to be in writing
7(a)	Advance Payment Bond	Certification by Employer to Surety of reduction to nil of Advance Payment
7(b)	Advance Payment Bond	Certification by Employer to Surety of dates Advance Payment is repaid
8	Advance Payment Bond	Bond is not transferable without written consent of Surety
8	Advance Payment Bond	Written consent not to be unreasonably withheld
9	Off-Site Materials Bond	All claims under the Bond to be in writing
9(a)	Off-Site Materials Bond	Certification by Employer to Surety of dates all listed items delivered
10	Off-Site Materials Bond	Bond is not transferable without written consent of Surety
10	Off-Site Materials Bond	Written consent not to be unreasonably withheld
VAT 1A.1	VAT Agreement	Effect of failure to give written notice required by cl.1A.2
VAT 1A.2	VAT Agreement	Contractor to give Employer notice of the rate of tax chargeable
VAT 1A.2	VAT Agreement	Amendment to rate of tax to be notified
VAT 1A.3	VAT Agreement	Rate given in written notice to be shown on each Interim Certificate

Clause No	Clause Title	Signpost
VAT 1A.4	VAT Agreement	Either party may give the other notice that cl.1A no longer applies
VAT 1.1	VAT Agreement	Provisional assessment of values
VAT 1.2.1	VAT Agreement	Action upon receipt of provisional assessment
VAT 1.2.2	VAT Agreement	Objection to the Contractor's assessment of values
VAT 1.2.2	VAT Agreement	Reply to notice of objection
VAT 1.3.1	VAT Agreement	Final statement of values
VAT 1.3.3	VAT Agreement	Employer to calculate final amount of tax upon receipt of final written statement
1.2	EDI	Changes to type of communication or persons operating system
1.3	EDI	Changes in technical standards etc.
1.4	EDI	EDI fulfils contractual requirements for communications to be in writing
1.4	EDI	Exceptions to previous item
2	Retention Bond	Notification by Employer of the date of issue of the next Interim Certificate after CPC
4(i)	Retention Bond	Any demand under cl.3 to be in writing
4(iv)	Retention Bond	14 days' notice of Contractor's liability for the amount demanded by the Employer
5	Retention Bond	Bond is not transferable without written consent of Surety
6	Retention Bond	Release of the Surety in the absence of a prior written demand

APPENDIX

Clause No	Clause Title	Signpost
10	Person in charge	PERSON IN CHARGE
23	Possession, completion	POSSESSION
26.2.5	Loss and expense	
5.3.1.2		
5.3.2	Contract Documents	PROGRAMME
5.5		
26.1		
26.1.1	Loss and expense	PROGRESS OF THE WORKS
26.4.1		